The refurbishment
of commercial and
industrial buildings

The refurbishment of commercial and industrial buildings

Paul Marsh

Construction Press
London and New York

Construction Press
an imprint of:
Longman Group Limited
Longman House, Burnt Mill, Harlow
Essex CM20 2JE, England
Associated companies throughout the world

*Published in the United States of America
by Longman Inc., New York*

© Construction Press, 1983

All rights reserved. No part of this publication may be reproduced, stored in retrieval system, or transmitted in any form or by any means, electronic, mechanical, photocopying, recording, or otherwise, without the prior permission of the Copyright owner.

First published 1983

British Library Cataloguing in Publication Data

Marsh, Paul
   The refurbishment of commercial and industrial
   buildings.
   1. Buildings—Repair and reconstruction
   I. Title
   725'.2'0286   TH4311

   ISBN-0-86095-030-1

Library of Congress Cataloging in Publication Data

Marsh, Paul Hugh.
   The refurbishment of commercial and industrial
   buildings.

   Bibliography: p.
   Includes index.
   1. Buildings – Remodeling. I. Title.
   TH3401.M37   1983   690'.52'0286   82-8177
   ISBN 0-86095-030-1   AACR2

Set in 9 pt Linotron 202 Times
Printed in Hong Kong by
Wing Tai Cheung Printing Co. Ltd.

Contents

Acknowledgements	vi
Preface	vii
Introduction	1

1. What is refurbishment?   3

Economy	4
The intrinsic value of the building shell	4
Conservation	4
Fortuitous reasons	5
Case study 1: Conversion of light industrial building to computer workshop	5
Case study 2: Conversion of bank premises to lettable offices	6
Case study 3: Conversion of fruit and vegetable market to shopping and recreational centre	9

2. Refurbishment: change of use or modernisation   10

3. The criteria of conversion   12

1. Survey report	12
2. Structural report	13
3. Services report	13
4. Discussion with statutory officers	14
5. Proposals	14
6. Costings	22

4. Fee construction alternative   23

Operational document	23
Case study 4: Upgrading department store	29

5. Practical problems of refurbishment – and their solution   31

Record drawings and survey techniques	31
Foundation reinforcement	33
Dampness	33
Wood restoration and care	39
Wall renovation and repair	42
Secondary walling elements	50
Roof renewal and upgrading	54
Internal structure remodelling	58
The serviced interior	61
Security and safety	64
Interior surface refurbishment	68

6. Examples of particular types of refurbishment   70

Case study 5: Theatrical upgrading	70
Case study 6: Industrial to cultural change of use	72
Case study 7: Refurbishment with occupants *in situ*	73
Case study 8: Conversion from domestic to commercial use	74
Case study 9: Disused factory to training centre	75

7. The handover   77

Maintenance manual: Lloyds Bank Building, 31–32 Park Row, Leeds	77
Estate agent's letting brochure	85
Appendix 1. Proprietary names and addresses	86
Appendix 2. Useful references and organisations	90
Index	91

Acknowledgements

The author wishes to thank all the manufacturers listed in Appendix 1 for the help, information and photographs provided by themselves or their public relations consultants.

Particular thanks are due to:

Abbey and Hanson Rowe and Partners for permission to reproduce photographs of the Lloyds Bank building project in Leeds and to quote extensively from the feasibility study report and maintenance manual.

Bovis Construction Ltd for permission to quote from the Operational Document for the Stephen Y project and photographs used of this and other projects.

Mr Malcolm Rickards of Rickards Timber Treatment Ltd for the loan of artwork for Figs. 5.15 and 5.16.

The Fire Research Station for permission to reproduce Figs. 5.37, 5.38, 5.39 and 5.40 (Crown copyright).

The Architectural Press Ltd for permission to reproduce Figs. 1.6, 6.5 and 6.6 from the Architect's Journal.

Preface

A very attractive book could be written dealing in broad terms with the refurbishment of prestige and architecturally precious buildings. Such a book may find a place on the coffee table, but not in the practical designer's or specifier's library.

Any book on building refurbishment, if it is to be useful to the person who is physically involved in the design or carrying out of refurbishment projects, must deal at the same time with generalities *and* with detail. It must relate how other people have tackled the sort of project that any practitioner might encounter; and then it must narrow the field to examine how some of the particular problems the project contained were resolved. In this way parallels may be struck and trains of thought triggered off which may suggest solutions to the reader's different, but possibly related problems.

Refurbishment could almost be said to be the art and practice of overcoming difficulties and making the most out of the raw materials (the existing building) with which the designer and builder are presented. Hence this book is about problems; often quite specific problems. They may not be your problems; but they could be very similar. If they are and ideas are activated, then the book has succeeded.

Introduction

All refurbishment projects are made up of a series of one-off problems, which have to be tackled by the designer on a largely *ad hoc* basis. Although some problems will be repetitive, common to several projects, the schemes as a whole will bear little similarity and each will have its own particular little clutch of difficulties, usually brought about by the type and structure of the original building and the new performance demands that are being imposed upon it as a result of the new use to which it is to be put.

In a way refurbishment doubles the complications experienced by designers of new building projects, due to the constraints imposed by the original structure. Refurbishment, therefore, presents a fascinating array of snags which are novel to the designer who has previously concentrated on new work. They produce a tangled maze that appeals to the problem-solver in us all, while frustrating the latent artist, who demands the *perfect* solution to every new building need.

The very diversity of refurbishment means that there is no way any book on the subject can deal in glib answers to every designer's problem. What it can do, however, is to point out the type of problems that can be encountered, give tips for their solution and, with the aid of a few case studies, show how certain particular cases have been treated. This precisely summarises this book's objectives.

As refurbishment becomes a more significant part of the total building activity of the developed world, so the special attitude of mind required by the refurbishment project will become a very necessary part of the building designer's glossary of skills. Not only will this ensure the designer's personal survival in the ever-increasing competition for building work, as new building projects diminish, but it will have other beneficial effects as well.

In some ways refurbishment is the antithesis of the 'disposable building' philosophy of the period which stretched from the mid sixties to the mid seventies. This attitude of mind traded in buildings constructed quickly for an instant need – and those that could be as instantly replaced by another limited-life building should the building user's needs change. We used to confuse this restlessness with progress. Had the phase continued we should have woken up to a land from which the whole visual consistency and development of building had been erased.

Conservation has now become a major influence in many aspects of our life and conservation of our building heritage has, therefore, begun to be appreciated as a desirable attitude. Apart from any aesthetic satisfaction we may derive from a consistency of national building style and appearance, the sheer wastefulness of tearing down everything that became older than thirty years is economic lunacy, particularly at a time when we are looking with concern at our dwindling natural resources. This general attitude has led to a more adult concern for the basic value of the products of the building industry, each one of which represents the amalgamation of a number of quite special skills, all at work in one of the most inclement industrial environments in the Western world – the building site.

The vast majority of these buildings were constructed of materials that would far exceed the practical life of the building. They, therefore, had a value in excess of the immediate need.

The refurbishment designer's job is to make use of this basic value and with it create a new life for the ageing building. The fact that he has as his raw materials an existing building is not necessarily a limitation and it may result in his producing a building of more charm and amenity than if he started from scratch on an open site.

It is hoped that this book will help the designer towards this end.

Chapter 1

What is refurbishment?

Refurbishment is in danger of becoming the fashion word of the 1980s, just as industrialised building became the fashion word of the 1960s. When fashion takes over, true meaning tends to become clouded and partisan philosophies obscure understanding. Such was the case with the industrialised building movement in the 1960s. Those involved in the movement were too concerned with its dogmas and their idealism to appreciate that already industrialised methods were beginning to be applied to conventional building, resulting in a confluence of the traditional and the revolutionary, which made a separate industrialised building identity unnecessary.

Today those who wish to preserve our building heritage have attached themselves to the idea of refurbishment, more for emotional reasons than reasons of economics. As a result there is a danger that refurbishment becomes thought of as synonymous with conservation. In fact refurbishment, although often embracing conservation, is much wider in scope and is thoroughly economically motivated.

It is true to say that refurbishment is to some extent an over-swing of the pendulum following the tear-it-down-and-build-something-new philosophy of the overheated 1960s building boom. It is also true that it is a reaction to the glossy and brash in contemporary design, which has its life in the lightweight, plastic and insubstantial. Refurbishment is, as it were, a flight back to the substantial and solid in building. But these are emotional boosts given to a building phenomenon which has a very sound and practical reason for its existence.

So what is refurbishment? Refurbishment should not be confused with conservation, although it does conserve the old and thus helps to preserve a continuous and evident building tradition. It is not, however, the overzealous preservation of outdated buildings which no longer have any relevance to contemporary needs. Only if a building has outstanding architectural merit can such preservation be justified.

Refurbishment is the hard-headed business of making use of what is usable in the ageing building stock; the skilful adaptation of a building shell (which is valuable in its own right and not due to any historic mystique) to a new, or an updated, version of its existing use. The existing building, once refurbished, should be equally as efficient in its new role as a purpose-designed building would be, given the usual number of restraints which always impede the designer realising the ideal in new or refurbished work alike. It is making use of what exists when it is useful. If a building has, by chance, some architectural merit and will, by its preservation, improve the amenity of the environment, so much the better. But that is not the primary *raison d'être* of refurbishment.

Refurbishment is also nothing to do with maintenance although, in the process of adapting a building shell for a revised use, maintenance will have to be carried out on the existing structure. But this is a secondary component of refurbishment and should not be confused with its primary purpose.

At the end of the day, refurbishment comes down to good management of the building stock of a country, a company or a building owner to ensure that the initial investment in 'bricks and mortar' is not squandered prematurely. We have seen enough examples of the way buildings of the past have been modified, extended and restyled to suit changing needs for this attitude of mind to be quite familiar and, therefore, require no justification.

Before examining the problems of refurbishment, let us look at the reasons why it is so relevant to the 1980s.

Economy

A hundred and fifty years ago building needs did not change as rapidly as they do today. The development spiral created by the Industrial Revolution had not accelerated to the pace of today. Building types fell into a few well-recognised categories and ones which changed very little over the years. But, with industrialisation, a restlessness reached into every corner of our lives, bringing with it different building types, closely related to the needs of the developing technology. Even our housing became quickly obsolete due to changing standards of comfort, ways of living and labour-saving aids.

It was this overheated development that reached its culmination in the almost completely 'disposable' society. All products were created to fulfill a particular need for a limited life span in the certain knowledge that, at the end of this time, the product would itself be out-of-date. This theory was applied totally to consumer products and even began to be applied to buildings.

Now a number of influences have conspired – some might say luckily – to make us pause a while and think again. The economic plight of the Western world is the most obvious cause, although this is only the symptom of our growing awareness of dwindling world resources – and particularly world fuel resources. At last it is beginning to dawn on us that the world cannot support this escalating rate of waste and we are being made to set a 'real' value on world riches.

Just as we are starting to consider the profligate way our staple resources are being frittered away and to look for means of reducing the drain on what is often a limited and non-regenerating store of material; just as we are learning to recycle our waste materials and get the most out of everything we use; so we are thinking of recycling our buildings to fulfill new needs for as long as their basic structure remains usable.

It is in this context that refurbishment has a definite part to play – and will continue to be a significant part of building activity for the foreseeable future. It makes good economic sense; and not only to the building owner, but nationally as well. In Case Study 1 an example of refurbishment initiated primarily by economic influences is given.

The intrinsic value of the building shell

Most buildings constructed of long-life materials (and most were so constructed until the 1950s and 1960s) have a basic structure that will outlast the majority of changing commercial and industrial needs. This means that, without substantial change, these buildings will in time fall into disuse or, at best, will be utilised in an ineffective manner; hindering rather than aiding the efficiency of the operations they enclose.

The contemporary habit of designing for future flexibility of use will assist the refurbishment of these buildings; but the large stock of buildings will predate this design attitude and may even be constructed of loadbearing walls, creating an inflexible cellular cross section. It is the reutilisation of this type of property that presents the most fascinating problems.

The impatience of the 1950s and 1960s often led to such buildings being demolished and replaced without a second thought. Today demolition itself has become an expensive operation and one that adds substantially to the overall cost of a replacement project. In addition, some of the recent techniques of construction (particularly in concrete) have presented the demolition contractor with an array of new and tricky problems which further add to the cost of demolition.

Case Study 2 presents an example of a project carried out largely because of the intrinsic value of the building shell and in spite of various structural and planning problems deriving from the original construction. This example is also used in Chapter 3, where the feasibility study, and Chapter 7, where the handover documents, are considered.

Conservation

It has already been pointed out that refurbishment should never be confused with conservation, although conservation is often the result of refurbishment. There are, however, a few – relatively rare – examples of buildings in which their value to the architectural, heritage or social history is such that the need to find a new use for the building is the main spur to refurbishment, and not the need to house a new function in an existing building.

Just such an example is that of the recent refurbishment of Covent Garden market by the GLC (Case Study 3). This project contains a considerable element of conservation, but is, nevertheless, true refurbishment in that the old market has been converted to fulfill a different – and very contemporary – need.

Other less prestigious examples will be referred to

Figure 1.1 Calder House exterior after refurbishment, Piccadilly, London

An extreme example of historic and aesthetic pressures resulting in the preservation of the external facade of a building, while the rest of the structure was completely demolished and rebuilt, is that of Calder House on the corner of Dover Street and Piccadilly (Fig. 1.1).

Here preservation of the two roadside facades was the main design and operational restraint. The contractors, Bovis Construction, carried out the work of protecting the facades, demolishing the structure behind and replacing it with a steel framed building containing 2250 m^2 of lettable office space on five floors, shops on the ground and mezzanine floors and a night-club in the basement.

This was more a cosmetic preservation than a true refurbishment contract. There are numerous examples of this type of total rebuilding behind a protected facade. Fig. 1.2 shows the forest of SGB scaffolding used to shore up the 18 m high external walls of the former Thames Tunnel Mills in Rotherhithe. Here this old five-storey flour mill in a conservation area about a mile from Tower Bridge has been converted into 71 flats with roof garden and communal facilities. Only the external walls and main cross walls were retained, together with the old water tower and chimney.

Fortuitous reasons

Refurbishment can, of course, be occasioned by a variety of major disasters, chief among which is fire. After large-scale damage has taken place, repair work is obviously called for. This only gains the status of refurbishment work when it becomes more than merely reinstating what was there before the disaster. If the opportunity is taken to upgrade the accommodation significantly, the project passes into the realms of refurbishment.

Refurbishment, therefore, is a very important and necessary part of the building heritage – today as well as in previous centuries. Just as the Gothic cathedral was added to and modified over the years, not preserving the less functional parts which previous generations of builders had provided, but moulding the building continuously into a form that matched contemporary needs; so today's buildings should be allowed to develop and progress. It is rarely justified to tear down what is only thirty years old, unless it contains some disastrous design or construction fault. Buildings should be allowed to live out their natural life span, being merely subjected to gentle alteration when the usefulness of their original form is past. This always, of course, assumes that the basic structure has intrinsic merit, structurally, functionally or architecturally.

Case Study 1

GenRad's Acoustic Vibration Analysis Division, European Headquarters, Maidenhead.
(Designers: Murdoch Design Associates)

The upgrading of a light industrial building and office to produce a computer workshop, office building and training centre.

Date of commencement of work: May 1979
Date of completion: November 1979
Total cost: £518,000

The existing building was typical of a large quantity of light industrial buildings in this country, usually between 50 and 20 years old, all drab sheds built on the edge of urban

Figure 1.2 SGB scaffolding shoring the external walls of Thames Tunnel Mills

throughout the text of buildings being converted to fulfill other uses, while preserving to a greater or lesser extent the original building for historic reasons or a desire to maintain the design character of a vicinity. The case of 31 and 32 Curzon Street, London W1 will be detailed in Chapter 6, when discussing adaptations of a domestic buildings to commercial use.

What is refurbishment?

Figure 1.3 GenRad Headquarters, Maidenhead; exterior

Figure 1.4 GenRad Headquarters, Maidenhead; interior

development. This particular building was about 20 years old. The American owner, GenRad, is said to be the world's largest manufacturer of automatic test equipment for digital, analogue and hybrid circuits. It wanted to centralise its European headquarter operations, previously undertaken in three cramped buildings in Bourne End, in this newly acquired complex at Maidenhead.

In spite of being unattractive in appearance, the existing buildings had inherent worth, particularly in their convenient location where permits for new light industrial buildings would then have been practically impossible to obtain.

There were three sections of the development:

- a single-storey steel portal frame workshop with brick walls, a few small windows, corrugated asbestos roof with translucent, corrugated plastic rooflights;
- a two-storey brick structure, originally a reception area; and
- at the rear of the site, a four-storey concrete frame structure with exposed concrete framework and timber infill panels.

This jumble of buildings had to be converted to house a high technology programming operation in which items of computer equipment built in the States were to be tailored to the individual needs of the company's UK customers, a training school for the customers' technicians and a European Headquarter offices. Clearly a thorough transformation would be necessary to achieve the type of accommodation GenRad would demand.

Murdoch Design Associates were commissioned to undertake the design work. It was agreed that the four-storey block was ideal for the office requirements, albeit situated rather inconveniently at the rear of the site. A new easily-identifiable glazed reception area was therefore attached to the four-storey block and the existing two-storey reception block became the training school.

The most significant aspect of this case study is the way in which this gloomy and shabby collection of buildings was transformed into a complex that has a light and airy character much more suitable to the highly sophisticated processes it was to contain. This was done at a considerably smaller cost than would have been required to build a new building. First the south-facing wall of the portal frame structure was replaced by a wall of bronze-tinted solar glass, changing the whole appearance of the building and forming an interesting contrast with the two multi-storey buildings it links. It also dramatically improves the internal environment making it totally unlike traditional industrial surroundings.

The building was reroofed with aluminium sheeting, hidden behind the glass elevation; while inside, a new power system was installed in an overhead grid, supporting the new general lighting fittings. Additional task lighting was provided to bring lighting levels up to 1000 or 1500 lux. A new warm air mechanical ventilation system was installed in the roof space above the grid.

Metal windows in the two multi-storey buildings were replaced with wooden windows, painted white, to provide a contrast to the dark brown overpainted concrete framework and brickwork. Internally these buildings were refitted with new partitioning and finishes.

This project is a good example of the way in which an unattractive, low-budget building can form the nucleus for an imaginative upgrading scheme, resulting in a pleasing external appearance and an interior fitting for its contemporary use and meeting the comfort norms of a progressive American company. More importantly the transformation has breathed new life into a depressing development which was in danger of slipping into premature dereliction.

The contract was undertaken on a fee construction basis by Bovis Construction and was completed in a mere 26 weeks.

Case Study 2

Lloyds Bank Building
Park Row, Leeds.
(Architects: Abbey and Hanson Rowe and Partners)

Conversion from bank and bank office accommodation to lettable office space.

Date of feasibility study (Chapter 3): September 1976
Date of commencement of work: September 1977
Date of completion: March 1979
Total cost: £566,896

The building was designed by Alfred Waterhouse RA (1830–1905), architect of the Manchester Town Hall, in an eclectic classical style. It is a Grade 3 Listed building which had been occupied by Brown's Bank in 1898, two years before its amalgamation with Lloyds Bank.

It consists of four floors and a basement giving an approximate total floor area of 3624 m^2 (of which 413 m^2 was in the basement). The walls are of brick; those fronting the adjoining roads being faced with grey/brown granite on the ground floor and alternating strips of brick and terra

Case study 2: Conversion of bank premises to lettable offices

Figure 1.5 Lloyds Bank Building, Park Row, Leeds; exterior

cotta above that level. The walls into the light well were faced in white glazed brick. At third floor level the facade above the cornice breaks up into much ornate detail, particularly on the pavilion at the north-east corner. The fourth floor is entirely in the roof space. The roof covering is generally Welsh slate with lead on the pavilion roofs (Fig. 1.5).

When the bank was about to vacate the premises, it was decided to commission a feasibility study to assess the viability of converting the building into lettable office space, upgraded to present-day standards. The feasibility study that resulted is quoted and discussed in considerable detail in Chapter 3. From this it can be seen that the architect's intention was to leave the external elevations as far as possible undisturbed – any alterations which were made being restricted to the walls of the light well at the rear of the property. This was in accord with the wishes of the Planning Authority.

While the building had a relatively low listing grade and demolition was not entirely out of the question, it was felt that if sufficient lettable office space could be achieved by refurbishment, the intrinsic physical and aesthetic value of the building shell were assets that should be preserved.

One of the major features of the refurbishment proposals was the extension of the mezzanine floor over the area previously occupied by the double storey banking hall. This produced additional floor area, giving a total building area of 3810 m², of which an estimated 2015 m² could be considered as lettable space on ground and upper floor

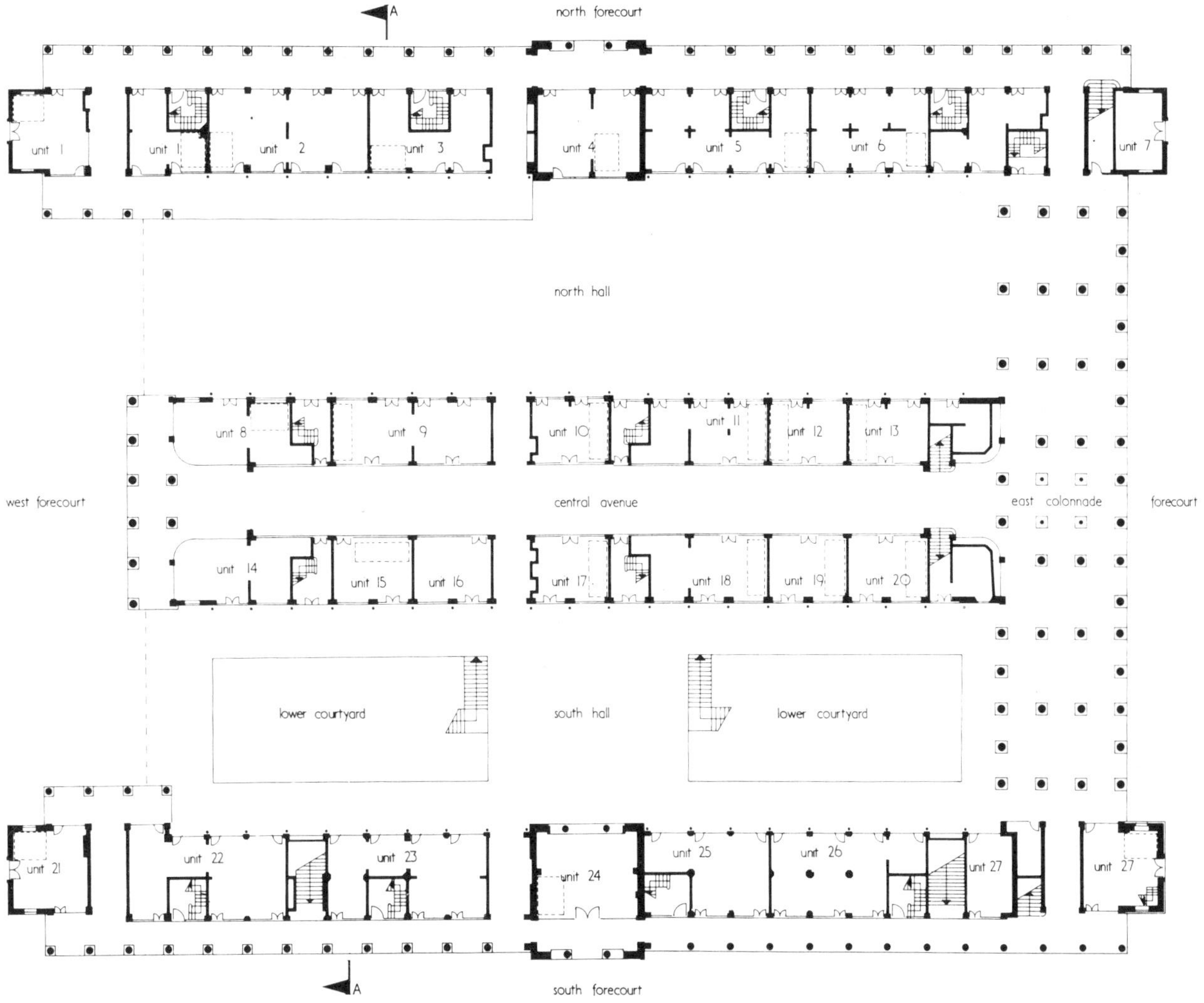

Figure 1.6 Ground floor plan of Covent Garden after conversion (courtesy The Architectural Press Ltd)

7

What is refurbishment?

Figure 1.7 Exterior collonade, Covent Garden

Figure 1.8 Southern hall, Covent Garden, showing the basement courtyards

levels. The basement was initially considered as a lettable restaurant, but the cost of conversion was judged too high to allow reasonable return and the basement area was instead let as storage space to the tenants of the floors above.

In addition to the enlarged floor area, the building was generally upgraded. Toilet facilities were increased and improved, an extra lift was installed and completely new engineering services were fitted throughout.

Case Study 3

Covent Garden Market,
Covent Garden, London.
(Architects: Historic Buildings Division, GLC)

Conversion of the fruit and vegetable market complex to speciality shopping and recreational centre.

Date of commencement of work: June 1977
Date of completion: June 1980
Total cost: £1,854,937

(The above statistics refer only to Phase III of the work and omit early exploratory costs and the restoration of the two cast iron roofs over the main halls)

Fowler's market building, designed in 1827, was a listed building. When it became obvious that the fruit and vegetable market had to be moved to an area which allowed space for expansion and with less congested communications – in fact, Nine Elms – it was hardly ever in doubt that the old market building should be preserved. The problem was what to do with it.

Should the building be preserved as a monument to Victorian England, or should it be put to work? The quality of the architecture scarcely qualified if for expensive mummification – and the cost involved, together with its ongoing expenses, could hardly be justified. In addition, the value of its site, both in terms of cost and the contribution it could make to the viability of that part of London, would have argued strongly in favour of the buildings being given some form of useful, present-day role to fulfill. To have resorted to mere mummification would have been a good example of conservation with very little social or community value. To find a vital and contemporary use for the buildings on this prestige site was essential. This raised the project into the classification of refurbishment; but refurbishment with heavy conservation overtones.

The major question, therefore, before the GLC was: what functions can the existing complex fulfill? To what use could this odd collection of two open-ended halls, four narrow lines of two-storey enclosures and a maze-like basement be put? (Fig. 1.6.)

After rejecting a proposal by the Architectural Association that it should take over the whole complex to form a new home for its school of architecture and provide meeting and social accommodation, it was decided that the development should never be allowed to fall into one ownership. Eventually the decision was made to convert it to house a number of speciality shops and restaurants, and provide permanent positions for small stallholders with a strong bias to craft and handicraft products. It was argued that this would breath new life into an area which had been in danger of becoming a dead spot in the heart of London. The open halls would be maintained substantially as public concourses which in time would generate their own social and leisure foci, as well as provide a site for the craft stallholders.

The main problem was how to provide enough lettable commercial space without impinging too greatly on the two halls. If insufficient revenue were generated, the project would be doomed financially from the outset.

The shop space above ground level was restricted to the Central Avenue, with its two opposite strips of double-storey enclosures, and the peripheral collonnades (Fig. 1.7). In order to boost the commercial space available, it became essential to bring into use some, at least, of the basement area, all of which was totally unusable in its existing form, due to the incredible difficulty of providing secondary means of escape. Even major alteration, if its use as a conventional basement shopping level was decided upon, would not solve the problem at anything but prodigious cost.

The solution that was agreed was to illustrate dramatically the difference between conservation and refurbishment.

Two linked courtyards were scooped out of the basement in the southern hall, forming two lower-level shopping arcades, complementing those at ground level – and which would house 22 small specialist shops – a total basement lettable area of 3000 m^2, or about 56 per cent of that in the whole building (Fig. 1.8).

It could be argued that the excavation of the two great holes in the floor of one of the halls was not in the true spirit of conservation and represented a major and unacceptable change in the original building. That it is a major change could hardly be disputed and is reason enough for this project to be classified as refurbishment and not conservation.

In fact, the unacceptability of this change can be refuted on the grounds that Fowler's Covent Garden market buildings had been continuously modified over the years as part of their normal and continuous use; now they were being converted to another commercial use while still preserving a proper architectural respect for the original buildings.

What has resulted in fact, now that the work is finished, is a lively concourse of street traders, pavement restaurants, live music and speciality shops which does more for the architecture of Fowler (as well as the livelihood of that part of London) than the occasional visit from parties of tourists to view a museum of Victorian commercial life could ever do.

Chapter 2

Refurbishment: change of use or modernisation

When a building changes its use completely it is clearly a case of refurbishment involving major structural and services modification. Such examples as the reuse of Covent Garden (Case Study 3) and the conversion of the derelict warehouse at Boston, Lincolnshire into the Sam Newsom Music Centre (Case Study 6, Ch. 6) are good examples of this type of refurbishment, with or without conservation overtones. The creation of a concert hall from the disused Maltings at Snape, Suffolk (Fig. 1.9) is a similar example of conversion from industrial to cultural use. However, these do not tell the whole refurbishment story.

Figure 1.9 The Maltings, Snape, after conversion

Often refurbishment takes place when there is no complete change of use. Maybe it can be a change of type of business within the same broad building classification that results in a refurbishment project. This took place in Case Study 1 where a 1980s form of light industry took over a building originally designed to house a different type of process and one which married up to the conventional view of light industry in the 1950s. A similar subtle change took place in the Leeds Bank building (Case Study 2). Generally the building had been originally designed as offices above ground floor level, but not offices of an acceptable standard today.

This, in fact, introduces the other extreme of refurbishment – the project in which the building in no way changes its use, but merely experiences a substantial modernisation programme. It is not uncommon for some old office blocks in city centres to be totally rebuilt behind their protected street facades, thus creating contemporary offices with a Victorian or Edwardian exterior. But this is an extreme example. Usually commercial and industrial modernisation consists of a thorough internal refit with some relatively modest structural additions or alterations – the sort of refurbishment that took place at the Theatre Royal, Nottingham (Case Study 5, Ch. 6) in which new structural work was restricted to new foyers and dressing rooms.

So what are the pressures that bring about the need for refurbishment? Ever since the Industrial Revolution there has been a changing attitude to working surroundings and conditions in industrial and commercial premises. Partly this can be due to technological development (changing industrial processes or more complex sophistication in office systems and equipment); partly to an increasing demand for more comfortable working conditions (first central heating, then ventilation, now air-conditioning – not to mention increasing space requirements); and occasionally to legislation leading to Statutory control – usually to assure the health and safety of the building's occupants. Fashion, too, of late, has become a not inconsiderable influence. These changing attitudes started to move slowly, but gradually increased in momentum as time went on.

Our Victorian ancestors showed admirable confidence in the rightness of their way of life. They constructed their commercial buildings in a form that did not envisage any future modifications. They did, however, leave behind a wealth of solid buildings that have considerable inherent value and can be converted to fulfil many present day needs (see Case Study 2). It has only been relatively recently that we have lost confidence in our vision of the future. Now we know we have not the vaguest clue what the occupation need will really be in a commercial building for longer than about 15 years – hence the design discipline of flexibility to allow easy modification to accommodate a changed need.

When refurbishment takes place as a result of a change of use, it is self-evident that the revamped building will fulfil all the conditions that apply to a purpose-made modern building. In fact the occupants of a refurbished building should be equally well accommodated as their opposite numbers in a new building. When, however, no such change of use is envisaged, it is likely that the industrial or commercial building may have to be thoroughly upgraded several times during its working life. A few of the major influences at work are:

- increased work space standards and comfort;
- greater complexity of equipment, plant and processes which require different accommodation;
- compliance with shops, offices and factory legislation and other statutory instruments;
- upgrading of fire safety requirements, either as a result of legislation or due to insurance pressure;
- modification of the structure to improve weather exclusion, acoustic or thermal performance;
- sheer commercial competition; either for staff or image.

The refurbished building should never be considered a second-best building. The ultimate aim of any upgrading project should be to provide its occupants with an interior of similar amenity and convenience to that in a new building – possibly a tall order on occasions, but a worthy aim nevertheless. In fact the refurbished building can in effect become a 'new' building – often the transformation of an ugly duckling of a structure into one that is pleasant to occupy and to look at. The story of Pilkington's training centre (Case Study 9; Ch. 6) is a good example of such a transformation.

The refurbishment designer who is presented with a building of architectural merit or vernacular quaintness, on which to practice his skill, is lucky indeed. He has it within his power to provide the revitalised building's occupants with comfortable accommodation in a building of architectural character that is of value to its surroundings.

Chapter 3

The criteria of conversion

Before any final decision is taken to convert an existing building to a new use or to upgrade a building to provide up-to-date facilities, a thorough investigation needs to be made into the condition of the existing building – its structure, the adequacy of its external envelope and its services, its usable floor area compared with the requirements of the brief, the scope for extending this floor area if necessary, the restraints placed on the development by its site, access restrictions and need for car parking, local authority or other mandatory curtailment, and the estimated cost of the work, maybe related to the revenue obtainable from the letting of space in the building, if it is to be so used in whole or in part.

All this information is usually compiled into a feasibility study which is prepared by a team of consultants, mostly under the leadership of the architect. The team is likely also to contain a structural engineer, services engineer, quantity surveyor and, maybe, a property expert, such as a letting agent.

As an example of the type of investigation which should be made and the feasibility report which should result, the case of the document prepared for Lloyds Bank Property Co. Limited for the conversion of the Lloyds Bank building in Park Row, Leeds into lettable office accommodation (Case Study 2, Ch. 1) is considered in detail.

A period of ten weeks was programmed for the feasibility study and the consultants' team was made up as follows:

Architects: Abbey and Hanson Rowe and Partners
Structural engineers: Ove Arup and Partners
Mechanical and electrical engineers: F. R. Jenks and Partners
Quantity surveyors: Northcroft, Neighbour and Nicholson
Letting agent: Adair, Davy and Mosley

The consultants' report will be reproduced, section by section, in its entirety – only omitting the final section relating to letting arrangements, which is irrelevant to the purposes of this book. With each section an explanatory commentary will be given to amplify the report and explain any apposite later events which occurred during the course of the contract.

Any initial study, however skilful, may fail to highlight facts which later, during the work, became evident and effected the carrying out of the work and its cost. The skill of a good feasibility study lies in the ability of each consultant to foresee possible areas of trouble and ensure that sufficient (but not overgenerous) allowance is made for such problems in the contingency sum.

1. Survey report

The building was surveyed in the light of existing drawings dated January 1961 supplied by the Bank Premises Department.

A visual examination of the whole building was made and check dimensions on each floor taken to ascertain the accuracy of the existing drawings. It was difficult to gain full access to the Basement because of Bank Strongroom equipment and for obvious security reasons. A more detailed examination is recommended when the equipment has been removed and the basic structure exposed.

The building fabric generally appeared to be in a sound condition but it was decided to carry out a more detailed structural examination and this Ove Arup's did on Friday 6 August.

The architects in this case were lucky to have a full set of record drawings. Often a relatively detailed survey has to be undertaken at this stage so that accurate sketch drawings can be produced.

Access to the basement obviously caused all the consultants difficulty. This is a normal problem when this type of study is carried out with the occupants *in situ*. The problem was aggravated here because of the building being used as a bank.

Generally the initial assessment of the building fabric being in a 'sound condition' was borne out by later events.

2. Structural Report

The only floor which we could readily open for examination was the fourth, which is somewhat thinner than the other suspended floors. However it is reasonable to assume a similar form of construction throughout.

The 4th floor is 11" (280 mm) thick as shown on bank drawing 1062M161 and 162 made up as follows:

1¾" (45 mm) Wood block
Nominal levelling screed
6" × 3" (152 × 76 mm) steel or iron joists at approx. 2' 2"
(660 mm) c/c surrounded by a "breeze"-type in situ concrete, with nominal top cover and 3¼" (80 mm) cover to underside
1¼" – 1½" (30 – 38 mm) coarse hair plaster.
¼" (6 mm) fine plaster

The joists span parallel to the facade. The concrete is similar in appearance, and possibly in strength, to breeze block, and the underlying plaster is providing considerable additional stiffness to the floor.

Should the wood block flooring be removed it would be necessary to provide a 25 mm concrete screed for fire resistance.

From this preliminary survey and our inspection of the premises the previous week, we consider that it would be feasible to remove the cross walls, although it may be necessary to leave piers at each end. The transfer of the floor loads from the existing walls to new beams would require careful and extensive propping. A system based on reinforced concrete rather than steel beams would be more suitable, as it would provide greater flexibility for dealing with hidden problems which may well come to light on such an old building.

We do not consider that the addition of a mezzanine floor over the banking hall would present any serious problems, provided that the foundations and brick vaulting are suitable, and we recommend a thorough inspection of the basements as soon as practicable. This would also establish whether or not the unevenness of the ground floor is due to settlement of the fill over the vaults.

Generally, as far as can be seen, the structural fabric of the building is in good condition and can continue to be used for load bearing purposes. However we consider that a thorough investigation of material types and strengths is essential if major alterations are planned.

Again access appears to have been a problem, limiting the extent of a structural investigation. Clearly it was vital to the feasibility of the scheme that the cellular form of the original plan, caused by loadbearing cross walls, had to be changed (see p. 16). The engineer pointed out the care that would be needed to prop the old filler joist floors when the cross walls were removed. The comments regarding the use of concrete beams, replacing the cross walls, indicate that hidden pitfalls were anticipated. These, in fact, were experienced during the work, resulting in part in an extension of the contract period.

Access problems again prohibited investigation of (and therefore comment on) the basement vaults. In fact these proved to be structurally sound, although some form of settlement had occurred over certain areas of the ground floor suggesting a consolidation of the fill over the vaults. A structural screed was used in these areas to level the floor to receive the new finish.

Note the comment about investigating material types and strengths prior to major alterations being undertaken. These tests were later carried out and considerable variation in the strength of the bricks was discovered. Generally, however, the walls were found to be adequate; the only strengthening taking place being as a result of the newly constructed brick piers, supporting the concrete beams which replaced the cross walls (Fig. 3.1), and the filling of disused flues with concrete.

Figure 3.1 Support structure after the removal of cross walls, Lloyds Bank Building, Leeds

Later exploratory work was undertaken, prior to tender, to establish more accurately the condition of such elements as the roof, the basement vaults and the stairs. In total the sum of about £2000 was expended on preliminary investigation work on the site.

3. Services report

The following is a summary from the initial report prepared by F. R. Jenks & Partners.

Heating and ventilating – general
In considering a change of use for the premises, involving major internal structural alterations, we would not recommend that any of the existing heating or ventilating plant be retained.

Piping and radiators are large and obtrusive with most piping exposed. The boilers will require replacing within five years and gas firing would be preferred. There is no significant ventilation at present and a ducted system on all floors would be required if windows are to be sealed units.

Electrical incoming supplies
The medium voltage distribution system at present supplied

to a basement room contains an assembly of eighteen switch fuses and supporting ancillary control gear. A 300 A incoming T.P. & N. switch fuse provides back-up protection to the busbars for all outgoing circuits.

This arrangement would prove to be unsatisfactory from a lettable tenancy point of view.

The Yorkshire Electricity Board would invoice the landlord with one consumer's electricity bill based on the appropriate tariff for the whole of the premises.

A vertical rising busbar main distribution system could be used, with sub-metering facilities provided at each floor level from a suitable central services duct.

On those floors measuring 80' × 96' a maximum of four distribution boards could be provided with an integral kWh meter suitable for monitoring the lighting and power consumption of each tenant.

In such cases where a minimal office space only is required by a tenant a pre-payment meter could be provided.

General lighting

A large proportion of the existing lighting is of the tungsten filament type. This greatly reduces the efficiency with a subsequent adverse effect on energy conservation.

Using fluorescent tube luminaries throughout will increase the light output considerably using less fittings, thus reducing capital expenditure and running costs.

As the tenantable office areas vary generally between 10' and 12', a recessed lighting installation could be included using polarised diffusers to minimise glare, especially where suspended ceilings are required.

Emergency lighting

The existing facilities available are provided by a 7 kVA portable generator used for the computer suites' essential supplies.

An emergency supply using a battery system for the strongroom area is also in existence.

In both these instances the amount of emergency illumination is inadequate to conform to the latest recommendations as suggested by BS 5266 Part 1 for the Emergency Lighting of Premises.

As space is usually at a premium and as the building would be used as lettable office space, we would recommend using a maintained mode of emergency escape lighting fittings strategically positioned at exits and public thoroughfares instead of a centralized battery system.

Small power and telephones

Dependent upon the size of the revised areas for lettable accommodation, we would recommend the use of twin compartment perimeter skirting to provide facilities for socket outlets, internal and external telephone systems.

Where floors may require to be screeded it would be advantageous to include underfloor trunking to integrate and supplement the perimeter trunking and thus enable greater flexibility of the power and telephone systems and reduce cable runs to a minimum.

Fire alarms

The existing fire alarm system, although satisfactory in principle, would not be acceptable for the envisaged new tenancy of the premises.

The Fire Precautions Act 1971 would require smoke and heat detectors to be included in certain areas.

Service and passenger lifts

Some of the lifts are very outdated and sluggish in response due to many years of service provided.

We would recommend the inclusion of two new passenger lifts in the centre core.

This summary is self explanatory. It is worth commenting that often the engineering services will be the part of the building which during refurbishment will have to be largely replaced. This results from the limited life of most engineering installations (approximately 15 years), the increasing demand for more comfortable living and working conditions, technical progress in plant and equipment (which has been considerable) and, now especially, the need to cut down building running costs and conserve energy. An efficient modern system of heating and lighting, although initially expensive to install, can dramatically reduce running costs.

Also the changed type of occupancy of the building can have effects on the services required (cf. the comment on fire alarms).

4. Discussion with statutory officers

(i) Planning Officer

Preliminary discussions ascertained the following:

(a) No application for change of use required.
(b) No alteration in designation of surrounding streets anticipated in near future (i.e. pedestrianisation)
(c) Increase in floor area with addition of mezzanine and toilets would result in an excess of the 10% max. permissible plot ratio. However because we are retaining an important building our request for additional area would meet with a sympathetic ear.
(d) No objection was raised to new building in existing light well providing external treatment is reasonable.
(e) Requested minimum or preferably no alteration to external appearance of building.

(ii) Fire Officer

(a) Would prefer enclosed fire escape route but will accept existing external fire escape stair providing access is maintained from both extremities of the office space.
(b) No dry riser or hose reels required. Loose fire fighting equipment to be agreed with individual tenants.
(c) Fire alarm system throughout. (See F. R. Jenks' report).
(d) Emergency lighting required on staircases and corridors. (See F. R. Jenks' report).
(e) Reserved comments on Basement pending layout of Restaurant and Kitchen, but was in favour of scheme as drawn in principle with number of means of escape as shown.

The drawings used in the discussions with the Planning and Fire Officers were broadly those attached to the feasibility study (Figs. 3.2 to 3.10), these forming the basis of the architectural proposals.

The Planning Officer's request for minimum (preferably no) alteration to the external appearance of the building should be noted. In fact this building is a Grade 3 listed building and the architect's scheme involved no alterations to the roadside elevations. Alterations in the light well were largely undertaken using reclaimed white glazed brick from elsewhere in the light well walls.

5. Proposals

(a) Architectural

The following constitutes an outline of the principles adopted in producing the scheme which is illustrated in the drawings included at the end of this report.

Proposals

Although no two floors are identical an effort has been made to produce a 'typical office floor' with maximum lettable space whilst taking into consideration the constraints of the existing fabric.

Floors 1–4

(a) *By removing as many 'cross walls' (walls at right angles to external wall) as structurally possible thereby creating the maximum number of options for one sub-internal division of the lettable space.*

(b) *By keeping the central service core as compact as possible whilst accepting that the existing staircase is rather wasteful in space but an integral part of the building.*

(c) *By building in the existing 'light well' to provide additional toilet accommodation (to satisfy Railway, Shops and Offices Act), together with attendant service duct facilities. This has the added advantage of tidying up the worst feature of the existing building.*

Main circulation routes have been maintained around the core giving the necessary access at two points to the fire escape. (Fire Officer's requirement).

All existing entrances and exits to the building have been retained with the Greek Street entrance affording a secondary access and exit to the offices above. The Russell Street entrance is ideal for an access point to the Basement as it is already 'half a flight' down due to the fall in the outside pavement levels. It also continues to provide a means of escape in case of fire as it links with the existing external fire escape at first floor level.

Basement

Has been opened out and planned to provide a basic shell without finishes to accommodate possible Licensed Dining

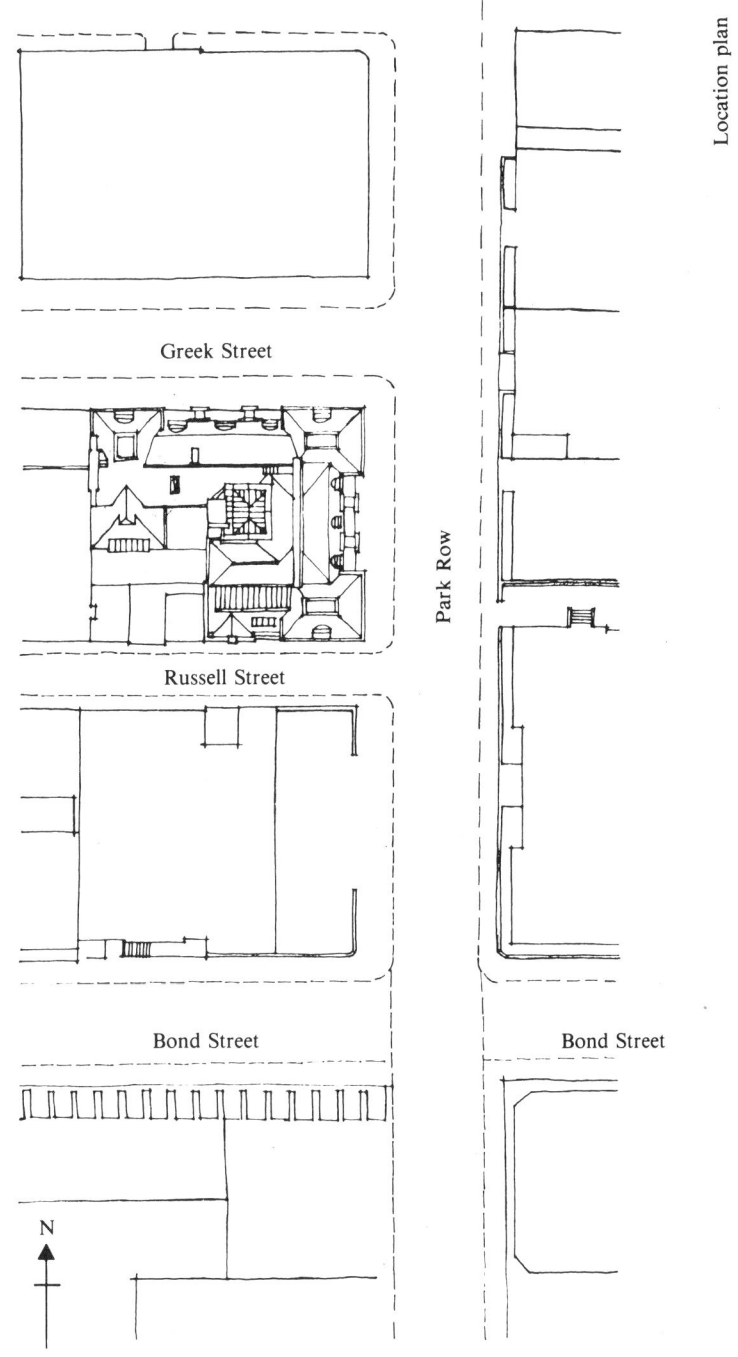

Figure 3.2 Lloyds Bank Building, Leeds: Block plan

The criteria of conversion

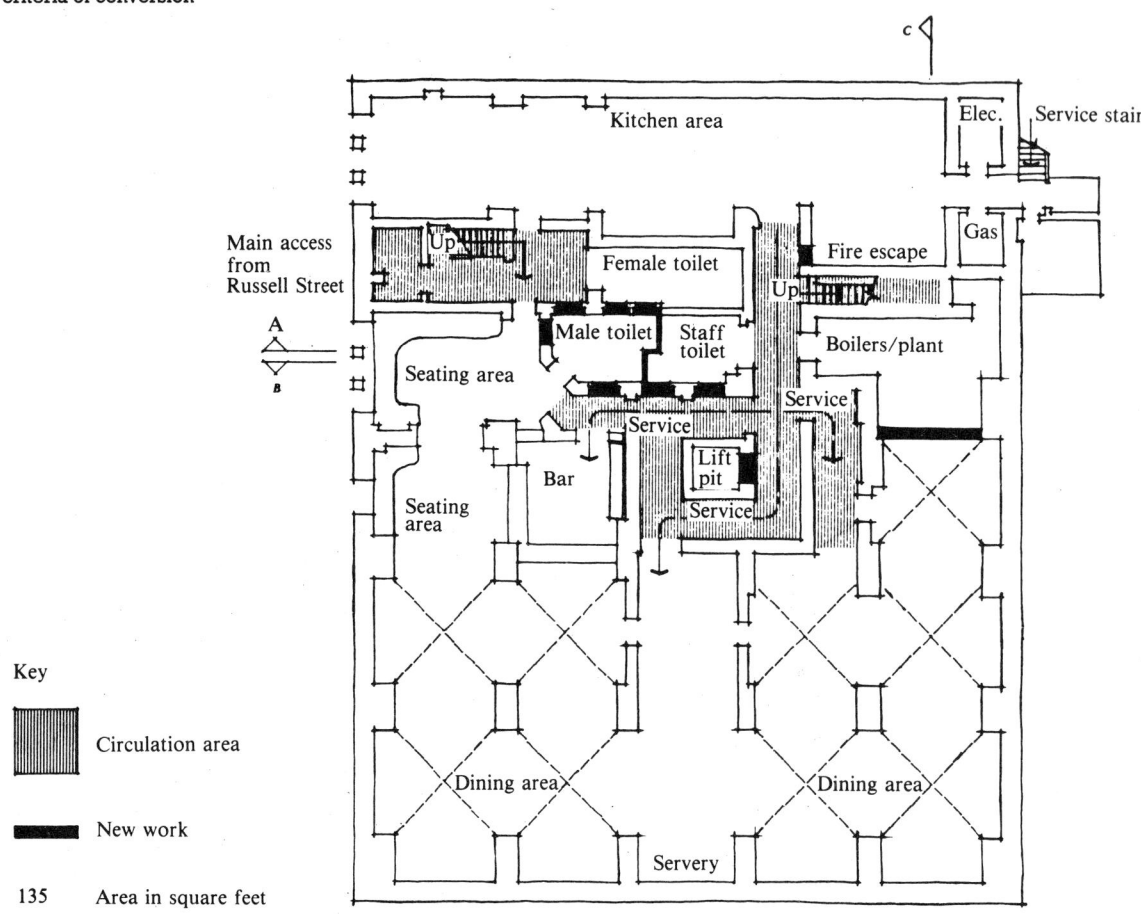

Figure 3.3 Lloyds Bank Building, Leeds: Basement plan

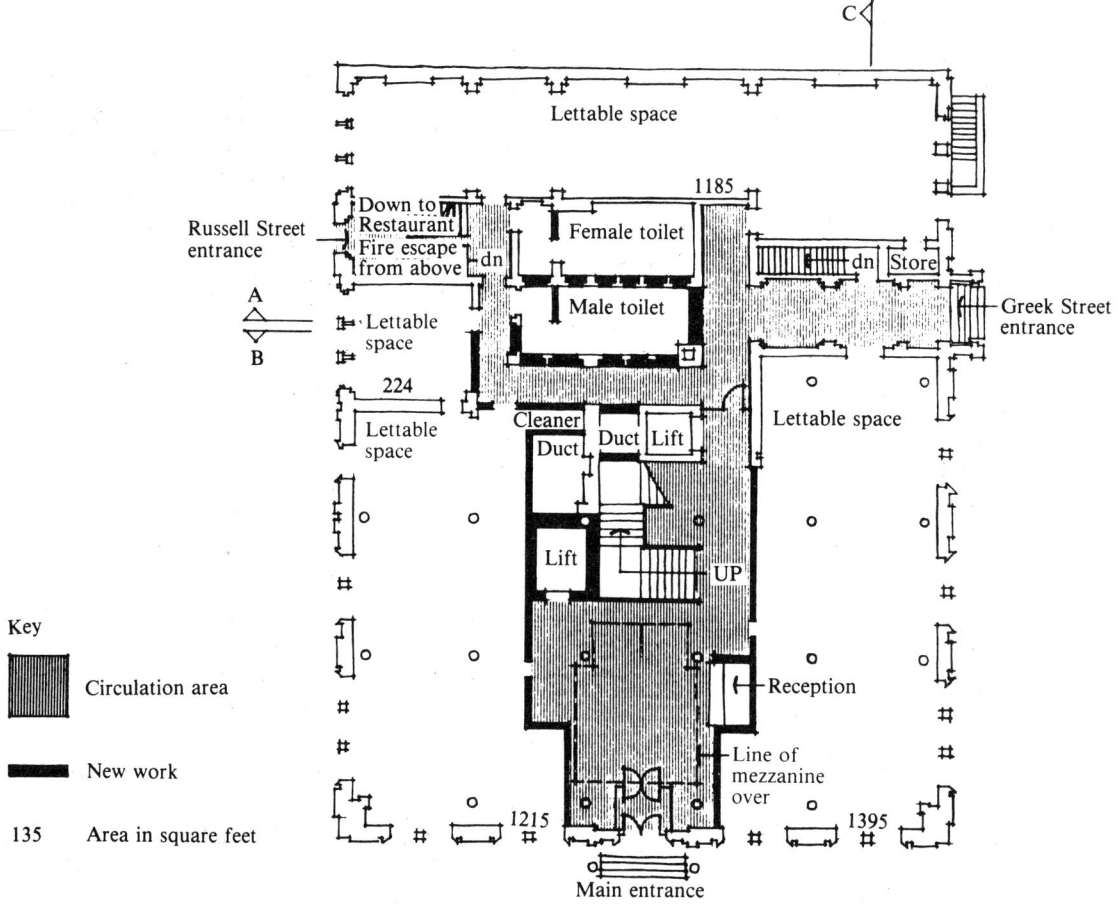

Figure 3.4 Lloyds Bank Building, Leeds: Ground floor plan

Services report

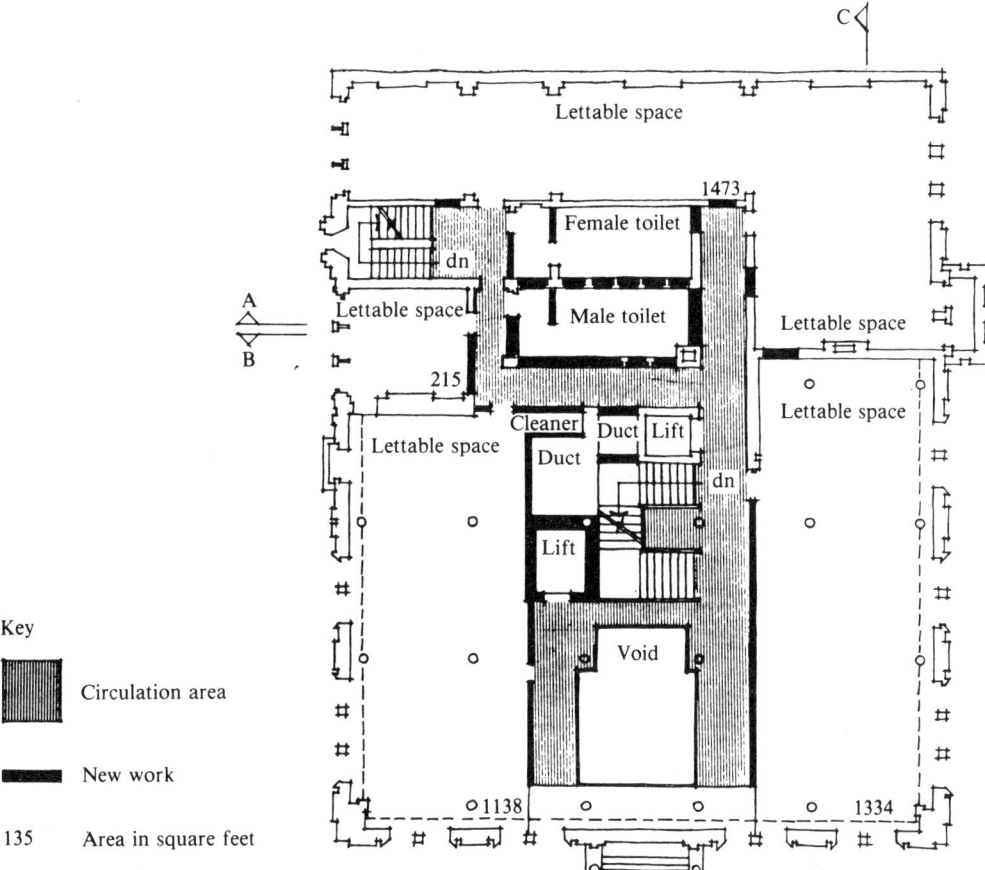

Figure 3.5 Lloyds Bank Building, Leeds: Mezzanine floor plan

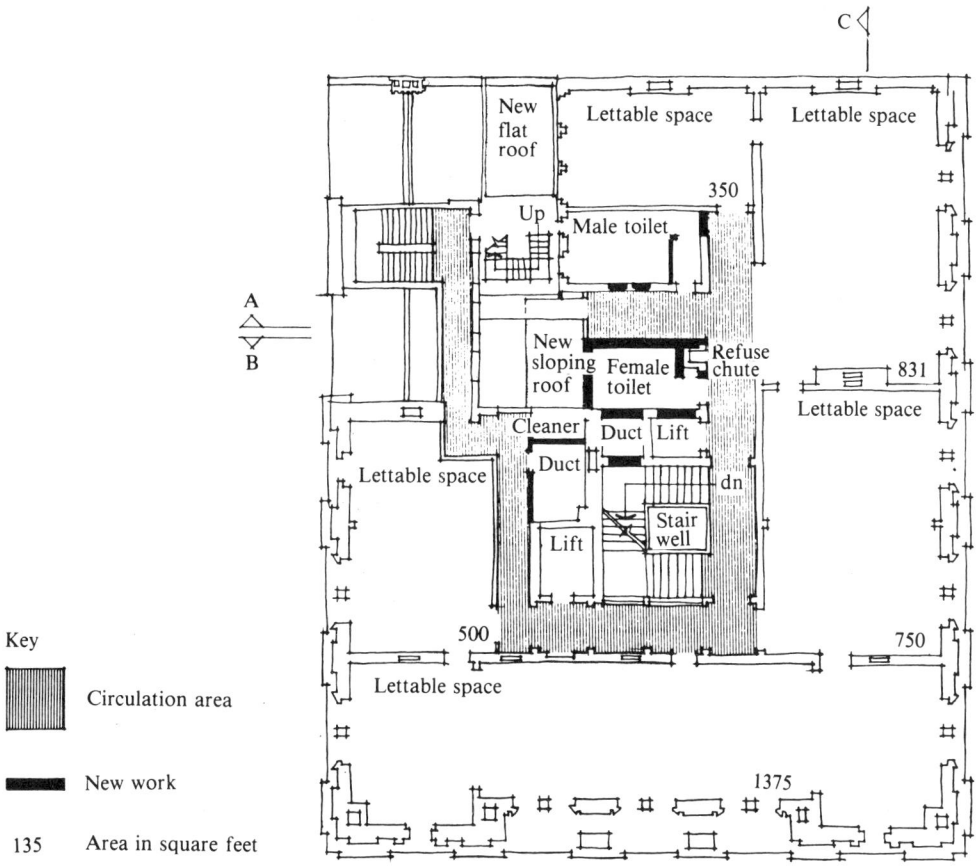

Figure 3.6 Lloyds Bank Building, Leeds: First floor plan

The criteria of conversion

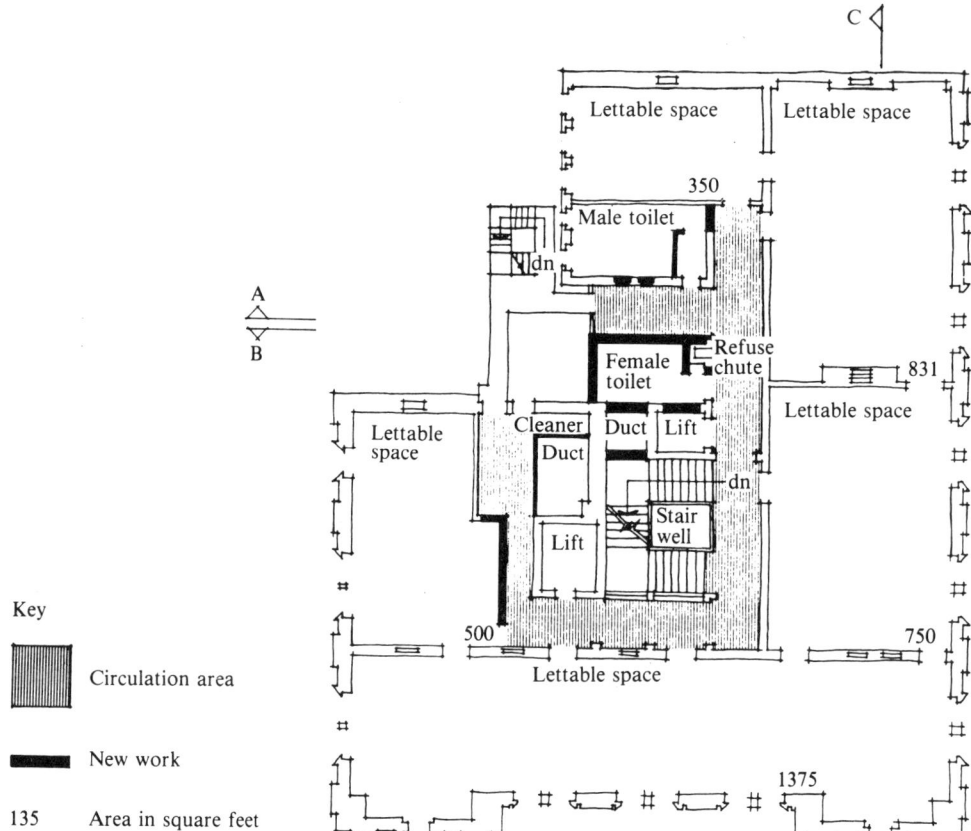

Figure 3.7 Lloyds Bank Building, Leeds: Second floor plan

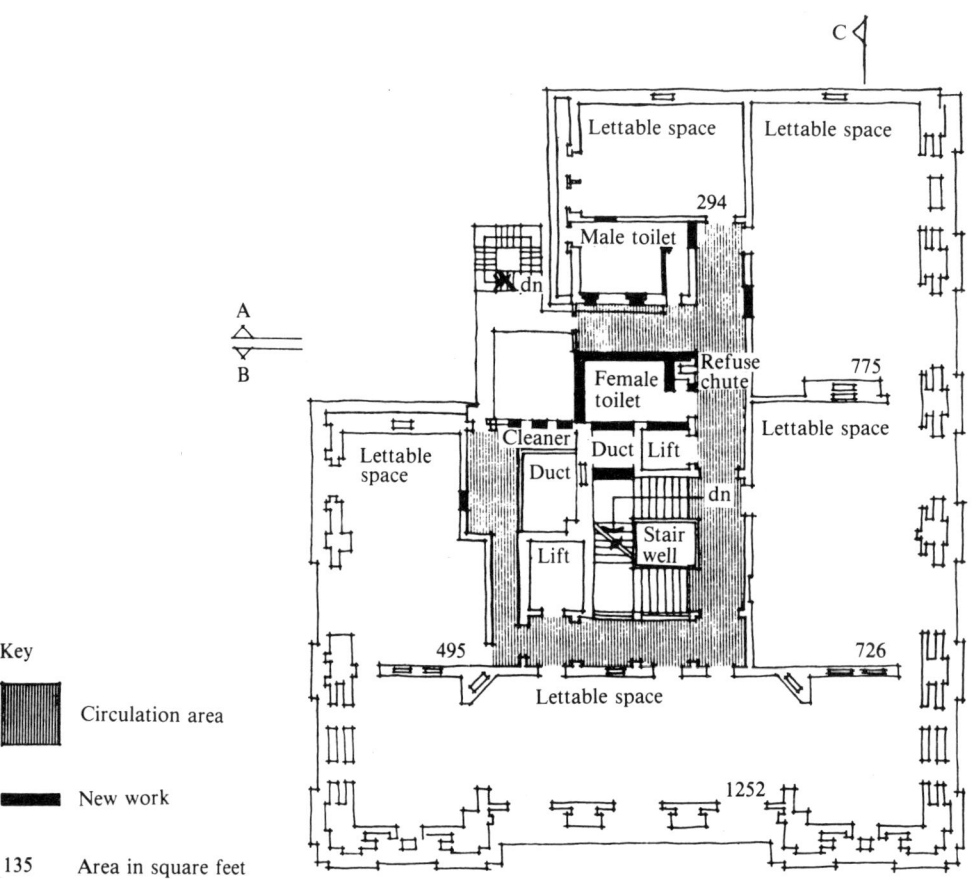

Figure 3.8 Lloyds Bank Building, Leeds: Third floor plan

Proposals

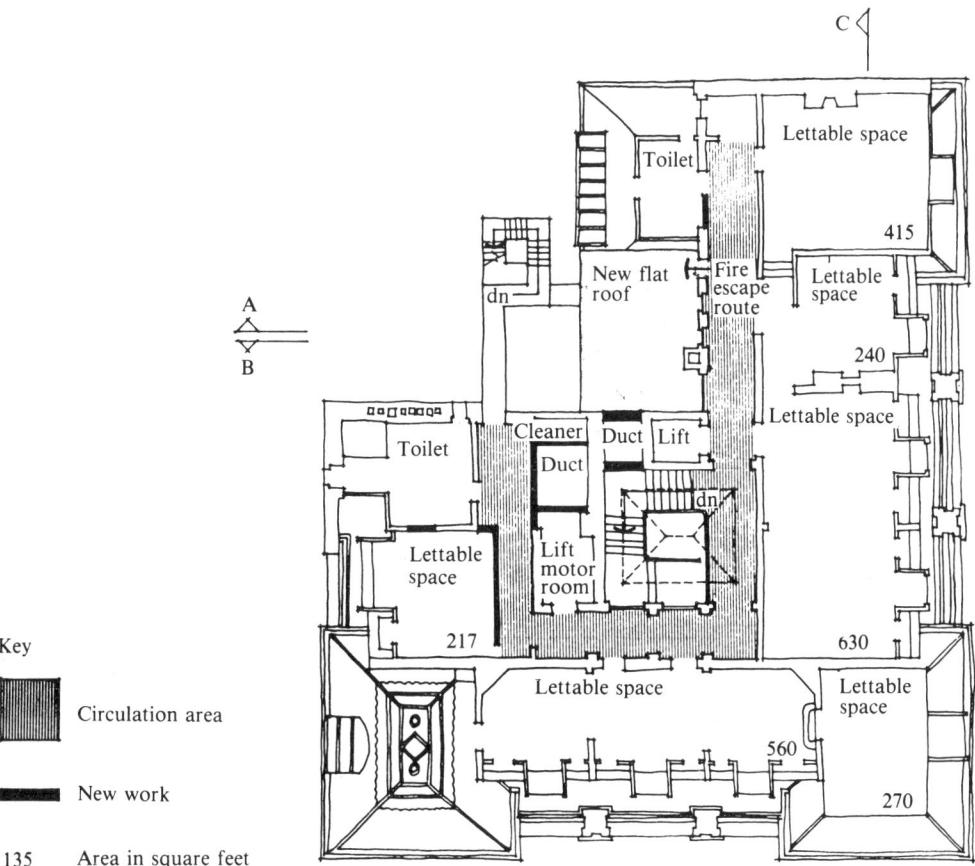

Figure 3.9 Lloyds Bank Building, Leeds: Fourth floor plan

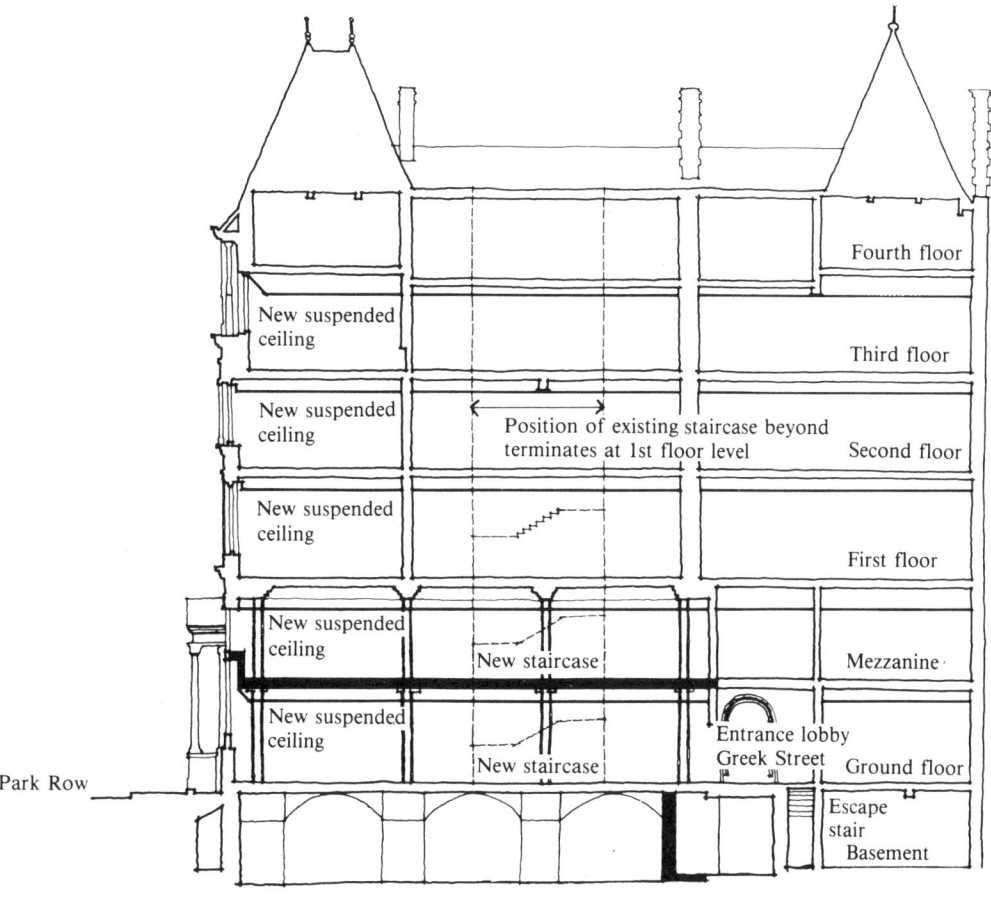

Figure 3.10 Lloyds Bank Building, Leeds: Section

facilities. The Planning Officer is obviously trying to encourage a Restaurant of this type in this particular area of the city centre.

Obviously the capital expenditure involved is high as can be seen from the quantity surveyors cost estimate and because of the fact that a reduced rental would have to be negotiated, or alternatively a sliding scale form of rental agreed, a number of options are available:

1. Provide scheme as drawn at high capital cost and allow tenant to 'fit out' the shell.
2. Carry out minimum structural alterations to provide adequate access and ensure future structural stability of foundations and fabric above ground, and then let tenant carry out all internal structural work plus fitting-out.
3. Retain as existing, only removing strongroom equipment and let as storage on a reduced rental.

The cost and letting potential are discussed more fully elsewhere in this report.

A point should be made that the letting potential of the fourth floor will be reduced by the plan configuration and changes in level. Also cill heights to windows, which are in the roof space, are high and natural lighting levels are not good.

First Floor
At this level the plan changes. The existing main staircase which changed direction to avoid penetrating the Banking Hall has been removed and continued down in the same position on plan as on the upper floors. It picks up the extended mezzanine and terminates at Ground Floor level in the new entrance lobby. Similarly a new lift shaft has been constructed, extending the old strongroom on the upper floors down to the ground floor. Together, with an upgraded car and remedial work to winding gear and mechanism on the existing lift it will provide adequate vertical circulation.

Ground Floor and Mezzanine
Are almost identical in plan with the original Caretaker's house completing the rectangle in the south west corner.

The existing mezzanine floor has been extended through to the Park Row frontage by inserting a new floor at the same level and occupying an area equivalent to that of the existing Banking Hall.

It would be possible to let these floors to a single tenant, and if necessary a separate staircase could be constructed linking the two. A new 'two-storey' entrance lobby has been formed using the original Banking Hall entrance from Park Row. A Reception point could be incorporated as shown if necessary.

Once more we see how crucial to the use of the floor area is the removal of the cross walls. These will be discussed in more detail in the next section.

The use of the basement was obviously an aspect of considerable discussion. Finally the matter was settled, as most of these matters ultimately are, on the basis of economics. It was judged that the return from letting the area as a restaurant would not warrant the expense involved in its conversion; and so the space was used simply as additional storage, related to the tenancies on the floors above. Little work was therefore carried out in the basement beyond tidying and cleaning the area and treating the external walls adjacent to Russell Street with waterproof rendering.

(b) Structural
The work involved in converting the structural fabric to provide the accommodation referred to in the previous section of the report can be broken down into two basic sections:

(a) Remedial work to existing fabric.
(b) New work.

(a) Existing
The method involved in stiffening existing floor slabs where cross walls have been removed is illustrated in Fig. 3.11.

We have assumed that the existing facade will carry the additional load placed on it consequent upon removal of the load bearing cross walls. This could be confirmed by the further materials investigation referred to in the initial survey report.

The other area of the existing building requiring attention is Ground Floor Slab, brick vaulting and foundations. This is an area which cannot be accurately assessed at the present time and if the project should go ahead further structural examination should take place under the supervision of the Structural Engineer. This would involve two men for one week approx. Because of the unknowns in this section of the work a substantial sum of money has been allowed to cover anything which might come to light when the tests are made.

(b) New Work
Mezzanine Slab – Flat concrete slab constructed to tie in with existing Mezzanine level with concrete upstand at perimeter and around void in entrance lobby. Existing columns to have all casing removed and new reinforced concrete casing cast round with 'dropped heads' to slab.

Toilet block in light well – load bearing external brick cavity walls with in situ concrete flat floor and roof slabs.

Later tests on the strength of the bricks discovered a wide variation in their crushing strengths. This led to rather larger piers being constructed to carry the new concrete beams which replaced the cross walls. The piers, it can be seen from Fig. 3.1 project considerably into the room and are the full width of the beam. A detail of the way the new beam 'stitched' together the two parts of filler joist floor is shown in Fig. 3.11.

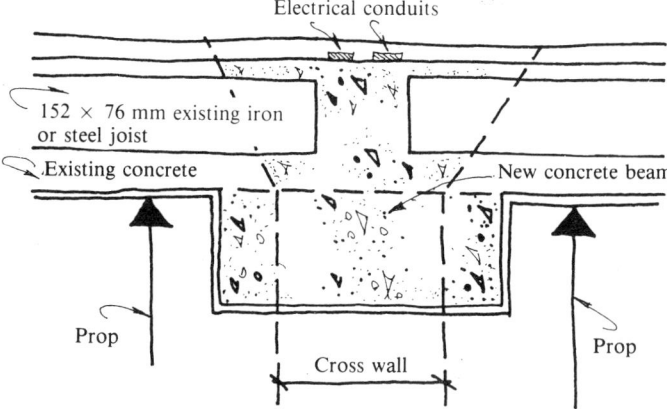

Figure 3.11 Diagram showing treatment of floor where cross wall is removed

After investigation it was ascertained that the brick vaults were sound, but subsidence had occurred in the filling. A structural screed was used to level these areas.

On removing the thin wooden lining that surrounded the columns in the existing banking hall, it was discovered that the steel columns had been encased in concrete, finished with attractive glazed tiles. A similar finish had been applied to the walls and once more had been hidden behind wooden linings. An effort was made to remove areas of this tiling, undamaged, so that it could be used as

Figure 3.12 New concrete columns supporting the new mezzanine floor

a feature in the new entrance hall. This, however, proved impossible.

The steel columns were exposed and were found to be cruciform in cross section, made up of steel angles. These were surrounded in reinforced concrete with mushroom heads to support the mezzanine floor (Fig. 3.12).

(c) Services

From the initial report it can be seen that the only sensible way is to pull out virtually all existing services and start from scratch.

Heating and Ventilation
An outline scheme has been prepared as shown on the overlay drawings related to typical floor plans.

General
The scheme provides a centralised boiler plant in the basement utilising two gas-fired boilers. Hot water is circulated by rising main to convectors, radiators and air handling units to all floors. Domestic hot water is provided by a calorifier situated in the Boiler room.

Basement
No services have been provided to the Basement with exception of the toilets.

Ground Floor and Mezzanine
Would be heated by finvectors around the perimeter and ventilation would be provided by plenum systems with extract return to a mixing chamber in the air handling units situated on each floor. The reason for air handling at this level is due to the pollution from traffic on Park Row, and as such tenants cannot be expected to open windows.

1st–4th Floors
Would be heated by radiators with no mechanical ventilation being provided.

Toilets
Heated by radiators with a mechanical extract system.

Circulation Areas
Heated by convector heaters plumbed to the hot water rising main.

The double flues from the new boilers could be brought up inside the building to 1st floor level and into the existing chimney stack which rises up above roof level.

Electrical

General
The existing switchroom at Basement level can be used to accommodate a new incoming switchboard or the existing one could possibly be modified.

A rising main busbar will be run from the Basement to the 4th floor in one of the ducts in the central service core. Distribution boards with sub-metering facilities being located at each level.

Basement
No lighting or powers provided at this level except in toilets although provision will be made at the main distribution point for the maximum load which could be expected if the area was let as a Restaurant.

Lettable Office Areas
Illumination levels have been designed at 500 lux using twin 85 W recessed fluorescents in areas where suspended ceilings occur and surface mounted fluorescents on a trunking system where existing plaster ceilings have been retained (mainly 4th floor). Surface mounted fittings will also be provided in circulation areas, staircases and toilets. A perimeter duct skirting trunking will be provided giving adequate coverage of socket outlets and telephone points. This will be linked to the central core and main rising busbar at points where cross walls have been taken out and conduit or floor trunking can easily be inserted. (See structural sketch for 'typical floor').

Fire Alarm System
The existing fire alarm indicator panel having 16 zones will be sufficient to meet Fire Officer's requirements. All fire alarm pushes will be linked back to the existing board.

Emergency Lighting
Will be provided to main circulation routes and at exits. A non-maintained form of emergency lighting is recommended.

Lifts
Two lifts are considered to be adequate to service the building.
Existing Lift – to be maintained serving Ground to 4th floor. Lift car to be upgraded and mechanism overhauled.
New Lift – to be provided serving Ground to 3rd floor with lift motor room at 4th floor level. Specification to BS 2655 10 person passenger lift – speed 200 ft/min – automatic doors having the facility for key operation so that furniture can be moved up and down the building.

Once more this part needs little amplification. It will be noted that the feed pipes to the finvectors can be seen on Fig. 3.1. Figure 3.13 shows how the area above the new reinforced concrete beams was used for the concealment of conduit and trunking. This area of floor was made good up to the level of the surrounding wood block flooring in a sand and cement screed.

Figure 3.13 Electric and communications cable distribution in floor

Schedule of finishes

Basement
No applied finishes with exception of toilets which will be identical to toilets on upper floors.

Lettable Floors (Ground to 4th)
Floors – Carpet tiles
Ceilings – Suspended accessible mineral fibre tile (*class 1 surface spread of flame*) with recessed fluorescent light fittings.
With exception of 4th floor. Existing ceiling to be retained on this floor. Lined with wood chip paper and painted 3 coats emulsion.
Walls – Patch plaster, line and paint 3 coats emulsion.
Skirtings, door frames and architraves – Paint semi-gloss.
Doors – New solid core h/w faced flush doors.

Circulation Areas (*including staircases*)
Floors – Vinyl tiles or sheet with nosings on stairs.
Ceilings – 3 coats vinyl emulsion.
Walls – 3 coats vinyl emulsion.
Skirtings, door frames and architraves – Paint semi-gloss.

Toilets
Floors – Vinyl tiles and skirtings.
Walls – Glazed ceramic tiles.
Ceilings – Plaster and paint 3 coats emulsion.

Exterior
Fire Escape and windows – Strip down prime and repaint gloss finish.

6. Costings

Estimated Total Cost	£452,000.00
Total Area (Gross)	41,000 ft^2
Cost per sq. ft	£11.02

The cost estimate is based on current rates and takes into account the present keen tendering climate.

A high proportion of this cost, approx. £68,000, is in work to the Basement area to convert it into a lettable shell, (*alternative 1, mentioned in the proposals*) and this figure does not include any services. The omission of the bulk of the work, say £50,000 worth, would have the effect of reducing the total cost to around £400,000 and the cost per square foot to under £10. The £18,000 being retained to act as a contingency against unknowns for structural work to Ground Floor Slab and below.

It should be emphasised that this is not an ideal way of carrying out the works from a practical point of view, especially if a tenant was eventually found for the basement restaurant and structural work had to be carried out at a later date.

There followed a report from the letting agent containing the estimated rental figures for the various floors and the likely return on the investment. This is largely irrelevant to the present book, beyond illustrating the obvious commercial basis for this type of refurbishment project.

The letting agent's brochure and the method of letting the building will be touched on in Chapter 7 which deals with the handing over of a refurbished building.

Chapter 4

Fee construction alternative

After the feasibility study, discussed in the last chapter, work on the project is likely to proceed in a conventional manner. Following the client's instruction to go ahead, the architect and engineering designers will start the production of working drawings, a Bill of Quantities will be drawn up by the quantity surveyor and competitive tenders will be sought. There is, however, an alternative method.

The fee construction method is highly appropriate to the project which, for commercial reasons, must start with all haste and complete as soon as possible. It is also particularly appropriate to refurbishment work which is often beset with uncertainties until the demolition work commences, and which is therefore difficult for the quantity surveyor to measure. In these circumstances the Bill of Quantities of necessity contains an abnormal number of provisional sums – and equivalent uncertainty in everyone's mind as to the ultimate cost of the project.

Fee construction allows the work to commence on site before the design is completed, it eliminates the need for drawing up a Bill of Quantities before tenders are sought and it also eliminates the costly and time-consuming business of going out to tender for the main contract.

All the work is managed by the fee contractor operating through a decision-making team (often called an operational team) including client, client's consultants (architectural, financial and engineering) and major subcontractors. All work is subcontracted and the contract expenditure is kept under continuous review by the fee contractor and the client's quantity surveyor.

It is theoretically possible to start work on site almost immediately the fee contractor is appointed and he has produced an initial estimate of cost and an *operational document*. A *cost control document* is produced to act as a basis for the contract to be financially monitored – and, when more detailed information is available, an *estimate of prime cost* is prepared by the fee contractor. Design work proceeds in accordance with an *information schedule*, produced by the fee contractor and included in the operational document. Because all participants in the work are represented on the operational team and have been selected because of their record of adherence to programmes of information production, it is said that the most cost-effective solutions to design problems can be achieved without impairing the programme.

As an example of the achievements that are made possible by the fee construction method, the refurbishment of 527–531 Oxford Street, London, is briefly set out in Case Study 4. This particular example is, maybe, an extreme one, due to the client's insistence on an early opening date for his new department store, prior to the Royal Wedding in July 1981. As a result of this, he was prepared to accept the cost risk of much of the early work which was carried out in advance of Statutory Approvals.

Operational document

In order to illustrate the way a fee contract is carried out, the operational document of the project described in Case Study 4 is examined in detail, the main body of the document being quoted in full with a complementary commentary on each section.

Fee construction alternative

The operational document contains a description of the works proposed, a statement of the methods to be employed to carry out the work, a construction programme conditional upon certain operational parameters and a programme of information requirements from the various consultants, essential to the achievement of the construction programme. This operational document establishes, with fair precision, the criteria upon which the estimate of construction cost has been produced.

Scope of work

Our proposal has been based upon Architect Drawings Nos. 1140/1,3,4, and 5 rev. C and additional verbal information given by Mr. Roberts of Murdoch Design Associates and Mr. Largesse of Andrews, Downie and Kelly.

In essence the works are assumed to be as noted below:

Strip Out
The stripping out consists of all existing relevant fitments, walls and partitions; ceilings including electrics; panelling; draining down of sprinklers and cutting off to suit new floor layouts, removing redundant staircases, etc. This has already commenced as instructed from Thursday, 26 March 1981 and will run concurrently to suit our proposals for new works.

Structural Alterations
Break out existing block walls in Basement between units 1 and 2 and 2 and 3.
Form void between Basement and First Floor including Escalator well and pits for Escalators.
Form new Fire Escape Stairs.
Form new Hoist Well.
Provide strengthening steelwork to support new services plant to roof.
Extend Mezzanine Floor in Unit 3 to provide total floor area. Also fill in staircases in Units 2 and 3.
Provide floor strengthening and steel trimmers for new brick or blockwalls, and services.
The full extent of alteration and propping details will have to be determined once Engineers details are issued.

Brickwork and Blockwork
New blockwork partitions are required to the basement plantrooms. The lift shaft is to be formed in brickwork. Remaining partitions will be generally to ½ h fire rating standard formed in metal stud and plasterboard.

Services
Comfort cooling is required for conditioned air, as opposed to a fully air conditioned system.
Existing Sprinkler services cannot be modified to suit FOC Regulations 29th Edition, and urgent discussions are to be held with the Fire Officer to establish that adaption to 28th Edition will suffice, or a new installation.
Provide new Electrical lighting and power requirements including Pelmet lighting to decor panels. Provision for track lighting to display areas, as required. Emergency lighting, as required.
Burglar Alarms and Public Address/Piped Music is to be incorporated.
Heating and Plumbing Services are required to Toilets and Offices together with necessary ventilation.
If necessary, modify existing services to suit new layout.
Provide Escalators and new Hoist including services.
A detailed services brief is to be agreed with Chapman, Bathurst and Andrews Weatherfoil.

Finishes
The Main Retail Areas: will have all walls fully covered in decor panels, either fabric wrapped or laminate finished incorporating a slotted strip system, wherever it is envisaged wall display could be used at some future date.
All columns to be mirrored on all sides.
Ceilings will be formed incorporating a tracked lighting system with recessed troughed lighting at 1500 mm centres. Services such as chilled air distribution and p.a. systems would be incorporated through grilles in lieu of lighting strips within the troughs. Sprinkler heads would be set out within the solid metal ceiling tiles. Subject to availability a 1200 × 300 mm ceiling panel would be used.
Perimeter pelmet lighting is required throughout.
The flooring is to be 80% carpet with terrazzo treatment to staircases, shopfront, and head and foot of escalators.

N.B. A fully terrazzoed floor has been considered but due to the time restriction this could not be incorporated. The programme would have been increased by six weeks.

The Offices, Stock Areas, Corridors, etc.: will be formed in lightweight metal stud and plasterboard construction generally to a ½ h fire rating standard.
Ceilings, where required, will be a lay-in grid fibrous ceiling tile with walls generally painted, but with glazed wall tiles to the toilets.

Shopfront
Specialist subcontractors have been approached to design and install the full height glazing system, as required. Toughened and laminated glass has been considered with stiffening fins at the entrance doors.

As can be seen, at this stage the contractor had received four sketch drawings from the architect, together with some supplementary verbal information. On the basis of this, Bovis had assessed the work content in overall terms which are summarised in this section of the document. By this stage the contractor may even have received competitive estimates from some of the major subcontractors – particularly the mechanical and electrical subcontractors – interviewed each upon its scheme to establish viability in respect of the overall programme and long-delivery equipment and selected the successful subcontractor. Long-delivery items may well have already been placed on order.

Site layout and access

Due to the frontage of the premises being Oxford Street all access will be from the rear of the site in North Row.
The exceptions to this will be the Escalator and the large plant such as Chillers which will have to be delivered on a Sunday morning in conjunction with police requirements.
Due to the fast nature of the project it is considered necessary to provide site offices, canteen and welfare accommodation at roof level provided GLC requirements are met i.e. 10 m away from adjacent property or 3 m if non combustible units used. Failing this, accommodation will be inside the building with two or three moves required as the project dictates. This will have a time and cost implication.
Because of the restriction, container skips will be used located in North Row or, if permitted by the Engineer, lorries may be used standing in the loading bay whilst loading, or a combination of both.
A Hoarding will be provided to the front of the building when the existing shopfront is removed. This will have to be relocated at a later date when the new shopfront is installed.
Once the new void is formed to the front of the building a hoist will be positioned for vertical distribution of materials.

The contractor in assessing the programme and costs has had to note the particular site difficulties associated with this project and may well have had initial discussions with the police and other official bodies to establish what restrictions are going to be placed on its activities.

Method statement

The extent of the works has been defined above and in the Architect's drawings.

Our programme referred to later and in this statement is based upon the scope of Works indicated. It is assumed that structural and service final designs will not indicate substantial changes to the scope. We will comment further under the section.

Working Hours
A day and night gang will need to be provided in order to achieve the 20 July 1981 handover.

Initially, the night gang will provide the necessary back-up to progress the structural alterations, i.e. scaffold erection, steelfixing, siting materials, rubbish removal, etc.

Immediately following the Easter holiday period this gang would need to attend to other trades such as services and associated finishing subcontractors.

Demolish and Clear Out
Clearance works are proceeding at present to clear out suspended ceilings, non-structural works etc.

Structural Alterations
Once the site clearance has taken place it is intended to commence on the structural alterations, commencing with the escalator well.

It is noted that at present it is planned to cross a beam at Ground Floor level. For the purpose of this appraisal we assume it will be possible to do this.

As there is an existing stair in Unit 1, but with a different disposition, it is thought the Engineer will require the new void and escalator trimmer to be taken back square to the fourth column.

In order to remove the concrete beams and pots a temporary scaffold would be provided under for safety reasons. In addition, where cross beams occur these would be broken out by compressor. If as a result of noise restrictions an alterative method has to be used, then either a thermic lance or diamond cutting would be utilised. It is assumed no support or strutting work would be necessary.

At Ground Floor level a solid concrete slab exists, and a temporary support scaffold with plywood on bearers would be fixed to allow the breaking out of the slab and void, again by compressor, or, failing this, by the methods outlined above.

Dependant on the trimmers required, it is anticipated that steel joists could be fixed to cleats which are bolted back by anchor bolts to the existing concrete. These would be encased by dry linings, except for anchor bearings of the escalator.

Existing staircases would be removed, with any necessary propping under structural members.

Openings for services, or the hoist, would be cut after new steel trimmers had been fixed underneath.

Until the structural design is further developed in line with latest layouts we are unable to finalise the requirements, but it is assumed that should major structural works involving frame and foundation redesign the layout could be revised to offset this possibility i.e. repositioning escalators, stairs, etc.

Brickwork and Blockwork would be carried out in the normal manner. These would be served from an electric mixer set up on the Ground Floor. Sand and mortar would have to be double handled from the street.

Services Installation
The basic premiss for the service installation is as follows:

Sprinklers
The existing sprinkler system does not and cannot be readily adapted to provide cover under FOC rules 29th Edition.

Design modifications are being prepared under 28th Edition for Fire Officer approval.

It should be noted that should approval be denied a new sprinkler system would undoubtedly delay the programme.

Comfort Cooling
Fan coils units would be located within the ceiling void and/or bulkheads, being fed by chiller water from the roof mounted chiller plant with condense lines running to RWPs.

Air handling plant would be sited in the basement plantroom feeding fresh air to the sales floor via grilles in ceiling bulkheads.

Extract would be taken over the void to roof level.

The basic heat for the trading area would be from the lights.

At the entrance a heated air curtain is to be provided.

Plumbing
Toilets are to be maintained in their existing position with new sanitary ware and W. C. partitions provided.

Electrical
Basically, this section divides into Lighting and Power, although the existing electrical distribution boards may require temporary provisions if serving other tenancies.

Basically lighting would be from recessed light fittings into a ceiling grid with track lighting for display. Pelmet lighting would be required to decor panels.

Power points to cash desks and island units would be from the ceiling, although a socket outlet would be provided at each column.

Emergency lighting would be provided with battery. It is assumed a generator is not required. Burglar alarms and public address systems would also be installed.

Deliveries
Due to the time span on the project it will be part of the proposed method to order plant before final details are agreed. This is in order to achieve the programme. The key items for installation are:

Fan coil units:	8–10 weeks
Chiller unit:	16–18 weeks
Air handling unit:	12–14 weeks
Boiler:	6– 8 weeks

Installation
The works will be phased into the programme in the traditional way although, if possible, there will be an overlap on the structural works on modifications to Sprinklers and of Plumbing works.

In general carcass work and wiring will be installed as soon as practical relative to the finishes, with sprinkler head levels adjusted. First and Second fix items will be carried out is conjunction with ceilings, decorations and decor panel work.

Finishes

As soon as practical suspended ceilings will commence particularly in the Sales area. As noted under Services we have assumed a lay-in grid type for speed of Services installation.

The extent of levelling floors will be ascertained and the screed, or existing screed flooring covered with a levelling compound for overcoming irregularities. No allowance has been made for taking up, or relaying a new screed.

Decorations to all services will be left as late as practical to minimise damage, although in a fast track situation it is found necessary to carry out the bulk areas to release following trades with some consequential making good necessary.

Shop fitting works are governed by the delivery of glass, and assuming that ceilings have to marry up to this the front should be fixed some six weeks before completion.

Toilet area finishes would commence with service installations as soon as the partitions are complete, followed by ceilings, decorations, floor finishes and w.c. partitions.

This section comprises a general statement of how the work is to be carried out. It is interesting to note that in the demolition clause, due to the urgency of the programme, work had already commenced on site. In fact much of the work was started before Statutory Approvals were obtained. This is referred to again under *Information requirements*. The contractor had, however, held discussions with the District Surveyor prior to embarking on the work and was reasonably assured of ultimate approval. The client was fully aware of the financial risks involved in this action, but considered the risks worthwhile due to his anxiety to open the store in advance of the Royal Wedding.

Another interesting point to note is that the structural design at this stage is by no means finalised. As a result the contractor assumes that the layout could be modified to avoid major structural complications.

The comment concerning the placing of orders on long-delivery items in advance of the agreement of final details is another example of the needs imposed by a very tight programme. In fact the availability of equipment played a large part in the selection of the equipment.

Construction programme and operational parameters

Programme FTP/1

Our programme is based upon a contract commencement of 6 April 1981 giving an overall completion by 27 July 1981 with handover by 20 July 1981.

In order to achieve this fast-track programme an instruction to proceed will be required on 6 April 1981.

A two-shift system has been described in our method statement as an essential requirement to achieve the programme.

The Easter holiday period has been shown as a break in production, but key operations would be progressed on site to advance essential building works.

The programme duration of 16 weeks (15 weeks to handover) is based upon information being released to suit the progress of the works. (See information schedules and notes).

As noted considerable overtime will have to be worked and early instructions will be required to order plant and materials before designs are fully developed.

A slightly extended programme OS/2 is attached indicating a slightly more economic process, for completion at the end of August 1981. An even more economic period would be to achieve completion at the end of October for Christmas Trading.

Subcontracts

It is also an essential prerequisite of the programme that subcontractors are selected who are capable of achieving the programme requirements.

Whilst competitive quotations will be obtained on an outline brief the subcontractor selected may not be the cheapest. On a fast-track project the right subcontractor is the key to achieving the programme. In addition with the outline brief given to those subcontractors precise pricing will have to be negotiated based upon further details, and pro rata to the budget figures obtained.

It would be the intention for the team to interview subcontractors, to assess their capability of achieving the programme, their knowledge of requirements and of possible alternative methods for time saving and cost saving, and their proven track record on similar projects. After this, we would then compare their budget figures and decide on a subcontractor to work with the team.

Instructions

Whilst it is normal procedure for Architects Instructions to be issued confirming drawings, instructions and verbal requirements on a fast-track project we propose that in conjunction with the normal system our job Memorandum Sheets ('Pinks') are adopted. In this way verbal instructions are noted and acted upon. It is a discipline that should be adopted by all team members. It should be noted that this form only records the instructions.

Cost Control

Once the Cost Control document is produced, initially, this will be the basis upon which the Contract is monitored. When further detail is available an Estimate of Prime Cost will be prepared.

Cost Controls relating to the EPC will be monitored and reported by the Quantity Surveyor and ourselves.

This section spells out the fee construction philosophy of subcontractor selection – that is, the capability of carrying out the work on time takes precedent over the cheapest price. It is also interesting to note the caveat regarding 'precise pricing' to be 'negotiated based upon further details, and pro rata to the budget figures obtained'. This encapsulates the idea of fee construction in which substantial work can be undertaken in advance of the final agreement of details while still maintaining strict financial control due to the 100 per cent participation of all members of the operational team and the 'open-book' contract accounting method employed. The almost invariable maintenance of the contract cost within the budget is evidence of the success of the method. The joint responsibility of Bovis and the client's quantity surveyor is set out in the last paragraph.

The programme that was bound into the operational document is illustrated in Fig. 4.1

Information requirements

Assuming a contract completion date of 20 July 1981 we have prepared a programme based upon known and assessed information.

It should be noted that when the survey is complete with the structural and service work determined and details issued, the periods indicated in our programme, and information release dates may have to be amended.

We have prepared a schedule of information that must be achieved if the overall date is to be met. It is appreciated that this imposes disciplines on all team members because of the manner in which information is released. Failure to achieve these dates will mean a delay to the contract.

Operational document

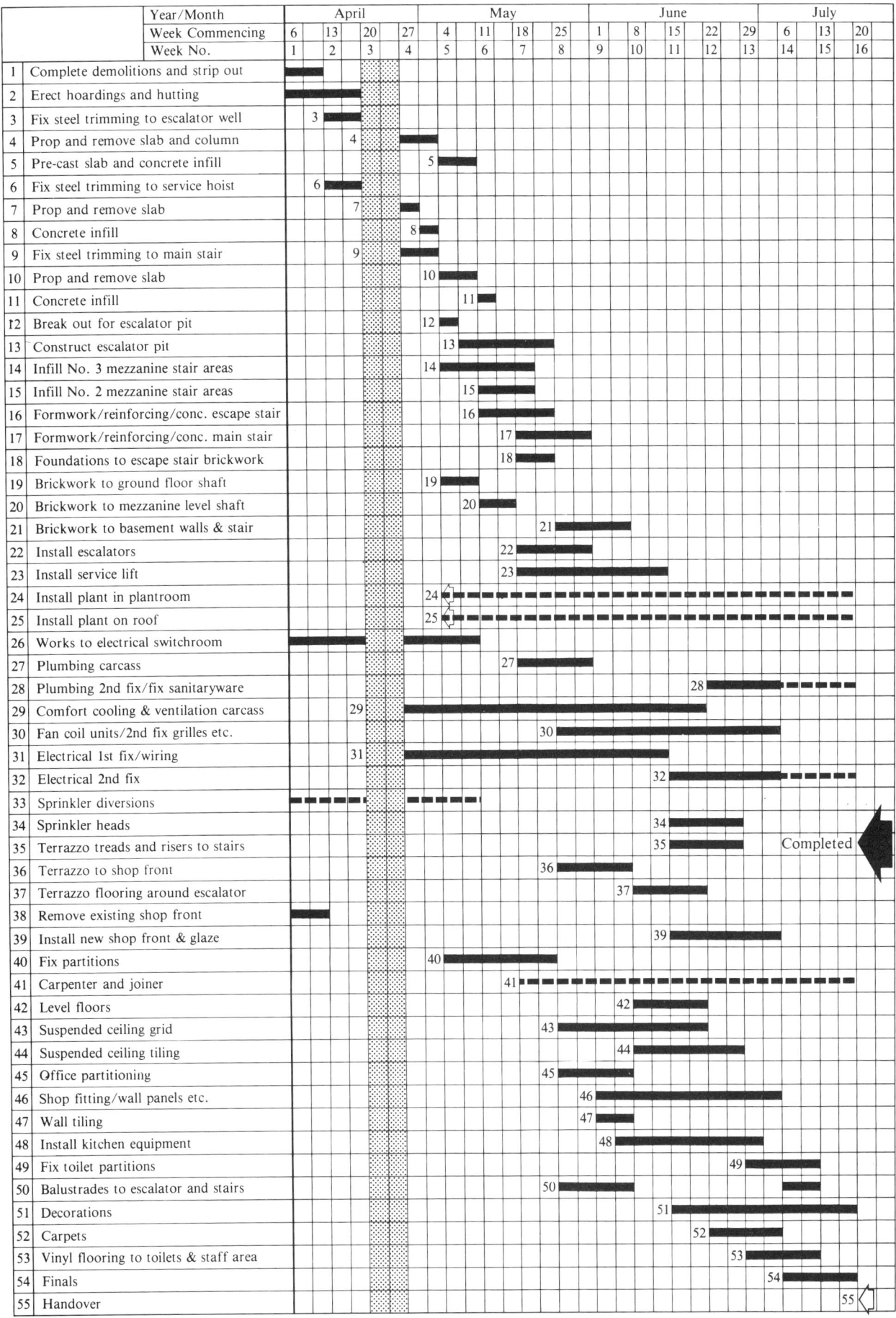

Figure 4.1 Fast-track programme; Stephen Y

Fee construction alternative

Information and drawing requirements schedule 527–531 Oxford St., London, W1.
Required from: Architect/designer (1)

Date: 25 March 1981
Amended: 27 March 1981

No	Item	Date required	Date received	Prog. start	Prog. finish	Remarks
1	Planning approval	26.3.81				Shopfront only. Or instruction to proceed without.
2	Building Regulations	26.3.81				
3	Fire Officers approval visit	26.3.81				
4	Insurance contingency	26.3.81				
5	Schedule of dilapidations	26.3.81				Done due for issue.
6	Agree datum grid			20.4.81		
7	Details of existing gas					
8	Details of existing meter					
9	Details of existing drainage	1.4.81				
10	Details of existing electricity					
11	Details of existing water					
12	District Surveyors approval	26.3.81				Or instruction to proceed without.
13	Environmental Health requirements	26.3.81				

Information and drawing requirements schedule 527–531 Oxford St., London, W1.
Required from: Architect/designer (2)

Date: 25 March 1981
Amended: 27 March 1981

No	Item	Date required	Date received	Prog. start	Prog. finish	Remarks
14	Party wall agreements	3.4.81				In hand, or instruction to proceed without.
15	Inform Local Authority Arch. dealing with planning.					
16	Dimensioned floor layouts	6.4.81	Construct drawings	13.4.81		
17	Required service including location.					
18	Required services including gas					
19	Required services including elect.	8.4.81		27.4.81		
20	Required services including G.P.O.					
21	Agree floor levels	4.5.81		1.6.81		
22	Requirement re Sub Station	30.3.81				
23	Plant plinths required	8.4.81		27.4.81		
24	Brick wall & block req'ts.	6.4.81		4.5.81		
25	Plastering requirements	13.4.81		4.5.81		
26	Matwell frames	4.5.81		1.6.81		
27	Concrete lintol schedule Engineer	6.4.81		20.4.81		

Figure 4.2 Information and drawing requirement schedule; Stephen Y

It will be noted that some dates indicate an immediate release date in order to obtain meaningful subcontracts. Also others such as those for ordering plant, escalators and shopfronts are in advance of the Statutory Approvals of the GLC consents and planning consents. In these cases the Client should be aware there is an element of RISK in preordering and proceeding. However, if the project is delayed until these are obtained the contract completion date would be delayed. It is obviously the Client's decision as to the action and RISKS he is prepared to take with some cost penalties.

Firstly it should be noted that the *Information and drawing requirements schedule*, included in the operational document, could be subject to amendment when the 'survey is complete with structural and service work determined and details issued'. The schedule, however, was all-important if the projected contract completion date was to be achieved and therefore all disciplines co-operating in the operational team were committed to this date and ways around any unanticipated obstacle had to be found if on-schedule completion was to be achieved.

The final paragraph in this section spells out the obligations placed on information producers (architectural or engineering) by the programme and explains that orders in advance of Statutory Approvals had already been placed in order to achieve the programme. The client's risk in these respects is indicated.

The first two pages of the *Information and drawing requirements schedule* (which was bound into the operational document) are reproduced in Fig. 4.2 to illustrate the detail with which information input is specified.

The fee construction method seems to have many advantages in refurbishment contracts, provided that the management expertise of the fee contractor is above question. This is particularly true where the quick reaction to a building need is important. It is a different way of working, but one that does not exclude any of the traditional personnel of a building contract, unlike some commercial package deal alternatives. These certainly have their place in refurbishment contracts, particularly in the case of some interior contractors, such as Ramchester or Project Interiors International, which demonstrate a considerable skill and design acumen. Where the contract involves more than interior conversion, with related mechanical and electrical engineering, it is possibly wise to choose between the conventional contracting method and the fee system. Where there is no time or opportunity to undertake all the necessary preliminary structural investigations that took place in Case Study 2, the advantage of the fee method, with its 'open-book' contract accounting combines financial control with quick reaction to a building need.

Case Study 4

Stephen Y Department Store,
527–531 Oxford Street, London W1

(Architects: Raymond Andrews and Partners)

The upgrading and refurbishment of three shop premises to create a fashion department store.

Date of commencement: 6 April 1981
Date of completion: 20 July 1981
Total cost: £1.2 million

Fee contractors: Bovis Construction Limited

A curious collection of retail stalls, known as Marbles Market and occupying the shells of three shop premises at the Marble Arch end of Oxford Street was the unlikely basis of the refurbishment that was to result in the up-market store called Stephen Y on three floors, specialising in ladies' and childrens' fashions, perfume and jewellery.

At the end of February 1981 the organisation that was to own the new store (Société Cysla d'Industrie et Commerce) contacted Bovis Construction to arrange a meeting. This took place on a Wednesday; by the following Friday an initial submission had been made by Bovis on the basis of the meeting and sketch plans produced by the client's architects, Raymond Andrews and Partners. The submission outlined proposals for a 'fast-track' programme to complete the work in 22 weeks. After a period of consideration the client gave his approval and demolition commenced only two days afterwards.

The owner of Société Cysla d'Industrie is a Mr Yeghiazarian, an Armenian who had been brought up to realise the benefit of commercial attack. At the age of 12, due to the untimely death of his father, he had begun trading on his own account from a mobile street stall in Beirut. In time he became the owner of a clothing factory, employing 400 people, and a five-floor department store. However, in the civil war in the Lebanon, the department store was destroyed and Mr Yeghiazarian came to London. Although he was starting once more to clamber up the commercial ladder, it was to be only six years before he owned his own 1700 m^2 Oxford Street department store.

A man of such commercial flair was unlikely to overlook the sales potential of the Royal Wedding at the end of July 1981. As a result, Bovis was asked to revise its programme to ensure completion in advance of the great day. Finally handover was agreed for 20 July, giving a sixteen week contract period – a tall order by any standard, but one which was further complicated by poor site access from North Row at the rear of the property, considerable structural alterations and the fact that a neighbouring shop shared the same heating system and had to be kept in business during the building works.

Major structural alterations included the clearance of internal walls which divided the three shops, cutting back of the first floor to allow the shop front to pass from ground level to fascia in uninterrupted sheets of glass, the formation of voids to house the escalators and new staircases from basement to first floor level and extensive temporary propping of the existing structure and the construction of new supporting structure. The proximity of the Central Line Tube tunnel was an additional complication.

In all, some 29 subcontractors were employed on site and up to 75 men were working at peak times on the project, often in day and night shifts.

Major subcontractors were appointed at the beginning of the contract. The fact that they joined the design/operational team at an early stage contributed considerably to the effectiveness of the programming. The whole contract was planned around the availability of the long-delivery items, like air handling and chiller units and the boiler. Corners were cut and much of the work was carried out at risk after preliminary consultation, but ahead of formal statutory approvals. The client believed the risk was worth taking.

The completed interior is a dazzle of mirrored walls and reflective ceilings, giving an illusion of vast internal space (Fig. 4.3). The floors are finished in high grade terrazzo and the glass shop front – said to be the largest single glass front in Oxford Street (5.4 m high by 19.5 m long) – is suspended from the fascia level to allow for the expansion and contraction of the building (Fig. 4.4).

Fee construction alternative

Figure 4.3 Interior of Stephen Y

Figure 4.4 Exterior of Stephen Y

Chapter 5

Practical problems of refurbishment – and their solution

This chapter is devoted to the examination of some common problems encountered in refurbishment and ways they can be overcome. Particular emphasis is placed on proprietary methods, material and services which may prove useful to the refurbishment designer. While this cannot hope to be a comprehensive review, it will give an indication of some available methods and, where possible, will give examples of refurbishment projects in which these methods have been used.

Appendix 1 gives the full names and addresses of all the companies mentioned in this chapter.

Record drawings and survey techniques

Before refurbishment design commences, an accurate record of the existing building needs to be made in the form of a full set of survey drawings of the building in plan, section and elevation. It is rarely the case that an adequate set of record drawings will be available; and if they do exist, it is likely that they will be in too poor a condition to be used for reproduction. They may even consist of a series of badly faded and torn dyeline prints, such as the left-hand part of Fig. 5.1. The right-hand section of the illustration shows

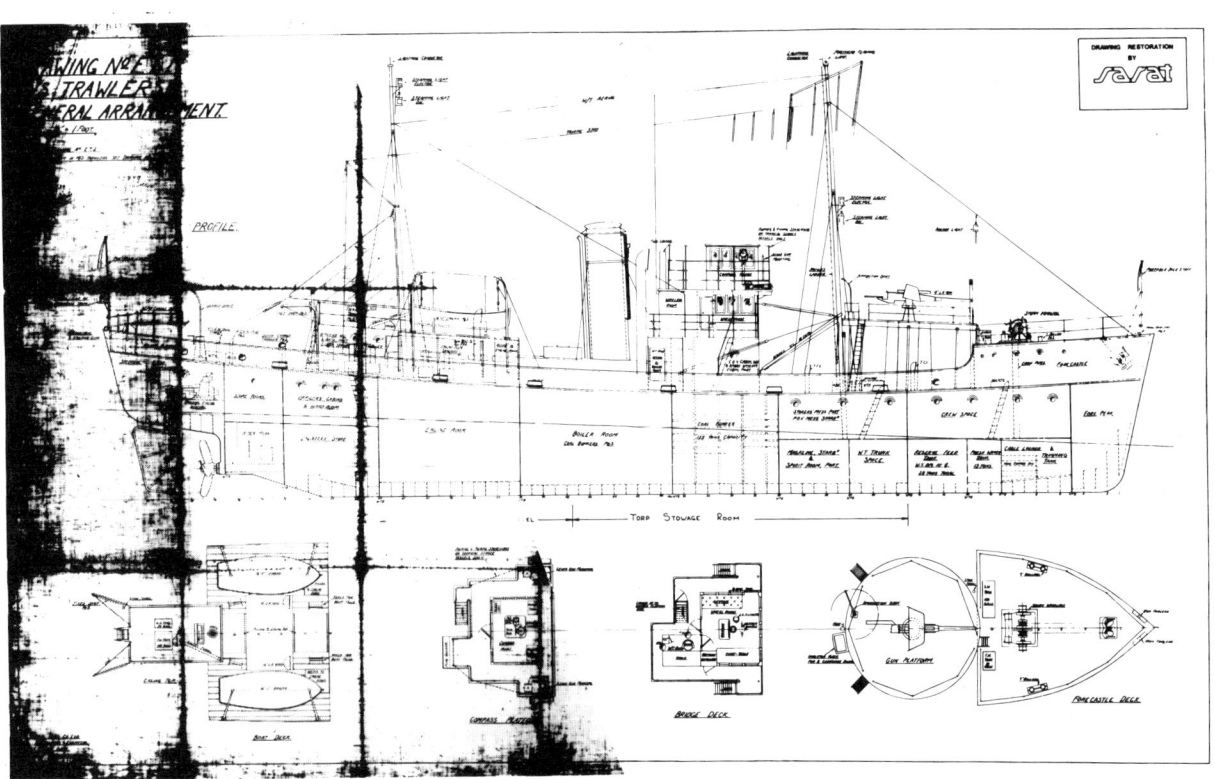

Figure 5.1 Print restoration by the Sarat technique

Practical problems of refurbishment — and their solution

Figure 5.2 Rectified computer elevational drawing

Foundation reinforcement

Underpinning of an existing building is a specialised subject and one that cannot be dealt with in detail in a book of this nature. Engineering advice should always be sought during the early stages of the feasibility study if there is any doubt about the structural adequacy of the existing foundations.

The need to underpin a structure can be the result of weakness in the original foundations, leading to differential settlement and cracking of the superstructure, or because of anticipated increased loading on the foundations as a result of the refurbishment. There are a number of companies who specialise in this type of work (for instance, Carrigan Underpin and Fondedile Foundations). They usually offer a complete survey, design and construct service and are prepared to guarantee their work.

A recent example of an underpinning contract, and one that is typical of many refurbishment projects, was carried out on old Port of London warehouses in the Cutlers Garden Development, EC2, by Fondedile Foundations. The 100-year-old loadbearing walls, constructed of substantial brickwork, showed no signs of structural distress. However, the refurbished building was going to impose increased loads on the foundations (a total load in excess of 1600 tonnes) and therefore it was decided that underpinning was essential and Pali Radice small diameter grouted piles were chosen for the work.

These piles are particularly suitable for refurbishment work. They can be installed without excavation and without cutting or shoring up the existing masonry. A series of holes are simply drilled through the existing footings and down to a satisfactory loadbearing strata. The piles are then grouted. They become an integral part of the existing foundations and, in effect, the existing structure acts as the pile cap. Piles can be set vertically or at an angle to provide resistance to lateral loads or overturning moments.

A further advantage of the system is that it is quicker (and therefore more economic) than many other systems. Its drilling equipment is very compact, thus being usable in restricted working conditions, and the work is carried out with little noise or vibration.

In all, 604 piles, each designed to support a working load of 25 tonnes, were installed at Cutlers Garden during a 15 week period (Fig. 5.4). The average pile length was 23 m and a maximum of 9 drilling rigs were used at the peak of the work.

A method of piling that is particularly appropriate to low rise buildings of a domestic scale is the Quickpile system of Ground Engineering. This method consists of pairs of 37 mm diameter pressure grouted reinforced concrete piles (one vertical, one raked) being installed at intervals along the walls. Each pair of piles is connected together by a reinforced concrete pile cap passing through the wall below the d.p.c. level.

A major advantage of this system is that all work can be carried out from the outside of the building with a minimum of disturbance to its occupants.

Dampness

Measuring dampness

Before considering various aspects of damp and its treatment in refurbishment schemes, a short note on the methods available for measuring the dampness of an object or structure is not out of place.

The most commonly used device for establishing moisture content is an electrical surface moisture meter of the

Figure 5.3 Original building from which Fig. 5.2 was made

how this drawing can be restored using a Photomanual drawing restoration system developed by Sarat (Process Photography).

This company specialises in various microreprographics, design and artwork and high precision photoreduction and printing services. Among its skills is this ability to restore old drawings, producing a new master negative on matt drafting or moisture-erasable wash-off film with inherent stability and archival permanence. The new masters are said to be fade, tear and deterioration resistant and give good reproduction quality. The restored drawing can also be changed in size during the process if required.

It is estimated that this service costs considerably less than conventional retracing or redrawing the material. It also has the advantage of more faithful reproduction of the original data than can be assured by the conventional method without extensive and careful checking.

A method of producing accurate and detailed line elevations from the building itself has been developed by Plowman Craven. This photogrammetric system is said to produce a record drawing of a building elevation in half the time and less than half the cost of a conventional measured drawing method. Suitable for all types of reconstruction work, elevations can be made to various levels of detail according to the requirements – from a basic outline of the structure to a detailed elevation showing every cracked brick or dropped lintel. Detail relating to structural deformation is also included.

Figure 5.2 shows a rectified computer drawing taken from photogrammetric prints made using a Wild Heerbrugg P31 camera. Figure 5.3 is a photograph of this typically ornate Victorian building.

These are just two examples of how present-day techniques can aid the preliminary tasks of collecting record information.

Figure 5.4 Pali Radice piling at Cutlers Garden Development

Protimeter type. In this equipment electrodes, pins or plates are pushed into or against the surface under test and a reading of the electrical resistance of the material taken on a 'relative' scale. This is not a direct reading of the percentage of moisture, but is a reading of changes in the electrical resistance that may be brought about by varying moisture content.

While this is a quick and simple method of establishing the presence of moisture, it is subject to certain drawbacks. A low reading is invariably a fair indication that dampness is not present; a high reading on the other hand need not necessarily indicate the reverse, as the presence of salts can inflate the reading. Since the readings are taken on or near the surface of the wall or material, they do not necessarily reflect the situation deep within the material and can be affected by surface dampness. Used with care and given sufficient interpretational skill, however, the equipment is invaluable. It is mobile, quick and simple to use.

A 'carbide' type of moisture meter ('Speedy' from Thomas Armstrong), is not liable to superficial inaccuracies, but is a much more laborious appliance to use. It involves the measurement of moisture in a sample core taken from the wall or material. Clearly this core can be taken from any position in the component being examined and is not influenced by local irregularities that may affect the surface dampness.

A laboratory method, involving the oven drying of samples taken from the component, is undoubtedly the most accurate method, but is most cumbersome and time-consuming. It is rarely necessary to use this method, except in the most problematic investigation.

Rising damp

It was not until the Public Health Act of 1875 that the incorporation of a d.p.c. approximately at ground floor level (or roughly 150 mm above external ground level) became compulsory in walls. Because most building materials are to some extent porous, with a capacity to soak up water in a similar fashion to a wick, many buildings that predate this Act suffer from rising damp. Indeed the same can be said of a considerable number of buildings constructed after that date; but this is because the d.p.c.s were (or became) inefficient due to the material from which they were made, the ageing of the material or bad workmanship.

Before leaping to the conclusion, however, that a d.p.c. has broken down, it is worth making sure that the rising damp is not due to some other cause, such as the bypassing of the d.p.c. by garden soil being piled against the wall, high levels of adjacent paving, bridging through external rendering, blocked cavities or other causes; some of which are illustrated in Fig. 5.5.

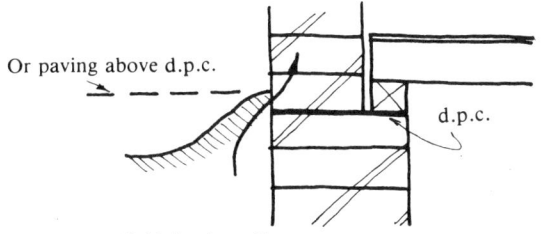

Bridging by soil

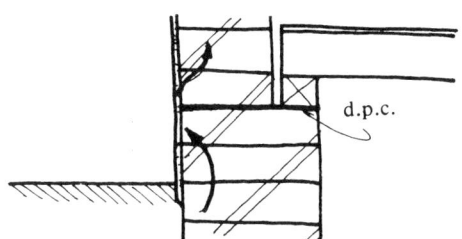

Bridging through rendering

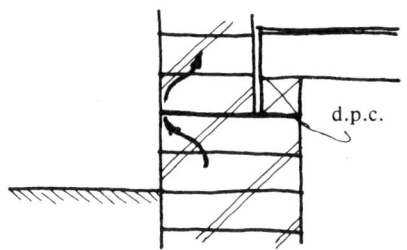

Bridging through pointing to d.p.c.

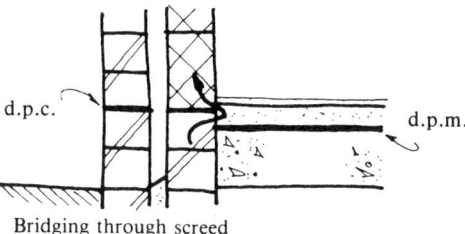

Bridging through screed

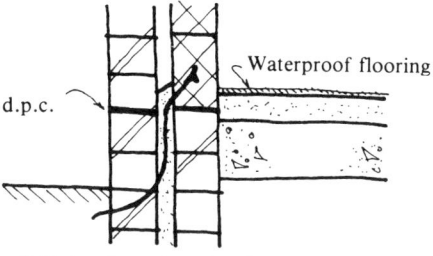

Bridging through cavity fill

Figure 5.5 Causes of rising damp other than from a fractured d.p.c.

One of the most difficult bridging situations to diagnose is that caused by a porous floor screed. There is no mandatory requirement for solid ground floors to have a damp proof membrane. Sometimes the floor finish is relied upon to provide a moisture barrier and unless the level of such a floor is exactly coincident with the d.p.c. in the walls, there is a danger of moisture bypassing the wall d.p.c. through the porous floor screed. A length of vertical d.p.c. to join the wall d.p.c. to the membrane or the floor finish may be necessary.

Rising damp can usually be identified by a roughly horizontal 'tide mark' on the wall at some distance above the floor – usually between 1 and 1.5 m high, the exact height depending on a number of factors such as pore structure of the walling material, amount of ground water and rate of evaporation from the wall. Above this tide mark no damage will be evident to the wall decoration; below, though, the plaster is likely to be disrupted and the decorations destroyed by efflorescence or bleaching, or finishes caused to lift and peel.

Care must be taken to distinguish between damp resulting from other causes, such as rain penetration, condensation or the presence of hygroscopic salts on the wall surface. The latter can appear on the surface of a wall for a variety of reasons. They may have even been deposited there by water ingress in one form or another and continue to produce apparently damp patches after the original cause of water ingress has been cured. Analysis of these salts can often help to diagnose the source of the dampness, if dampness truly still exists.

Salts analysis Because ground water invariably contains soluble salts from the soil which concentrate on the wall surface as the dampness evaporates, an analysis of these salts can give an accurate indication if the dampness is due to rising damp. For instance any water passing through a wall may contain salts dissolved from the walling material – chloride from unwashed or dirty sand and sulphates from cement and other building materials. The ground usually contains both of these groups of salts, together with nitrate which is necessary for plant growth. If, therefore, the salts on the internal wall surface contain nitrate, it is a fair indication that rising damp is present. On the other hand, if sulphates only are present it is extremely unlikely that the damp is rising from the ground. The same applies if chlorides only are present.

Most companies that specialise in damp remedial activity can undertake this type of test; as also they can test for the hygroscopicity of the salts present. This is an important factor, because some salts, which may have been deposited on the wall surface, are likely to be hygroscopic and attract atmospheric moisture under conditions of 75 per cent (or greater) relative humidity, thus producing damp patches even after the original cause of dampness has been removed. If hygroscopic salts are present, clearly replastering has to take place. Usually this involves replastering an area to a height at least 300 to 500 mm above the tide-mark.

Measures to remedy rising damp will depend on the cost and estimated effectiveness of the treatment when compared with the value, condition and life of the property. If the building is being refurbished, it is likely that a fair expenditure is worthwhile; but do first ensure that relatively cheap remedial measures, such as reducing the level of surrounding earth, will not solve the problem before resorting to the expensive business of inserting a d.p.c.

When it is established that the real cause of the rising damp is a non-existent or fractured d.p.c., the available options are:

(a) inserting a new d.p.c.
(b) draining and drying the wall
(c) concealing the damp.

Inserting a new d.p.c. This course of action should lead to a complete and permanent cure of the defect. It is also a potentially expensive remedial action. It can either be effected by physically cutting out and inserting a barrier to the rising damp, or by injecting a chemical damp-proof course, or by using a system based on electro-osmosis.

Physical membrane insertion The traditional method of inserting a d.p.c. is to cut out a course of bricks, a short length at a time, and replace it with engineering bricks or slates. A less expensive method is to cut into the bed joint at an appropriate height in the wall and insert a membrane as the work proceeds. Generally a length of about 0.5 m is treated at any one time, but in the case of heavily loaded sections (at jambs and junctions) shorter lengths may be more advisable.

The material used as a d.p.c. will vary depending on the depth of the slot which is cut into the joint. This in turn will depend on the type of mortar and the saw used. (Any Building Centre can advise on the type of saw which the BRE recommend for this use.) The common characteristics of a d.p.c., however, are impermeability and durability. The d.p.c. should project about 5 mm on each side of the wall; in the case of a narrow slot, a thin rigid sheet material is required (half hard copper or zinc, if only a 10–15 year life is expected); in a wider slot two layers of a thicker material (bituminous felt laid to break joint) will be required to fill the slot, or a thin sheet (copper or polythene) spread with a bed of mortar to act as a packing with slate wedging slips at 0.3 m centres can be used. In the latter case, soft copper is suitable, but high or low-density polythene is a cheaper alternative. High-density polythene is more easily inserted and all types of polythene should be filled with carbon black to retard deterioration.

Generally the treatment is time-consuming. The rate of progress will depend on the state of the wall and the type of saw used. Where the joints are of dry crumbly mortar in regularly coursed brickwork, the rate may exceed 3 m per hour in a 220 mm wall. Where there are course irregularities, the rate may be depressed to 0.6 m per hour. On an average, rates seem to be in the order of 1.3 m per hour in 220 mm walls and 2 m per hour in 105 mm walls, where the walls are of an average standard.

Chemical d.p.c. An injected chemical d.p.c. usually involves the injection of a water repellent into holes drilled at regular intervals along the base of the wall. This is a difficult process because the walls usually treated are themselves nearly saturated. Consequently the formation of a continuous barrier is difficult. The most successful techniques appear to be those employing siliconate solutions in water or a siliconate/latex mixture. These are miscible with water when injected, but later change character to form a water repellent band. Siliconate solution is usually allowed to permeate under the influence of gravity; siliconate/latex solution is forced into the wall under pressure. Operative care and attention to detail is vital in either case (Figs. 5.6 and 5.7).

There are many proprietary systems available on the UK market, some of which are included in Appendix 1. Only methods having an Agrément Certificate should be considered. The beauty of the method is that it can be used in very thick walls where the installation of a physical membrane would prove very difficult, or even impossible. Drilling patterns may vary with the system used, but usually each brick or stone in the course treated is drilled at about 115 mm centres and to a depth about two thirds of the wall thickness. Thicker walls can be treated from both sides. Where high density brick is used, drilling can be in two adjacent bed joints and at 125 mm maximum centres.

Such a case of very thick walls was the 17th Century Old Brew House at Castle Bromwich, which had walls up to 560 mm thick. There was about 60 m run of external wall which needed treatment, together with several lengths of internal wall. *Peter Cox*, using its patented diffusion process, was asked to treat the problem. The Peter Cox method allows silicone fluid to permeate by osmosis throughout the entire thickness of the walls. Savings of up to 50 per cent on traditional methods are claimed and the system is covered by an Agrément certificate and guaranteed for 30 years. The insertion of frozen sticks of silicone solution at 150 mm centres is the basis of the Norman Rudd Freezteq Silicon d.p.c., and is said to work well in rubble infill walls, unlike many chemical d.p.c. systems.

Electro-osmosis The scientific basis of these systems has been the subject of controversy. There are several fundamental types in existence, all of which prevent the rise of significant amounts of water above a certain level in a wall by electrical means. In one system electrodes of similar metal connected together are placed, one in the wall, one in the ground. In another an electrical potential from an external source is applied between such electrodes. In yet another, both sets of electrodes are placed in the wall, but at different heights in the wall – the outside electrode being lower than the inside electrode. Dissimilar metal electrodes are used in another case – one in the wall, the other in the ground – galvanic action producing an electrical potential.

It has been claimed that systems employing dissimilar

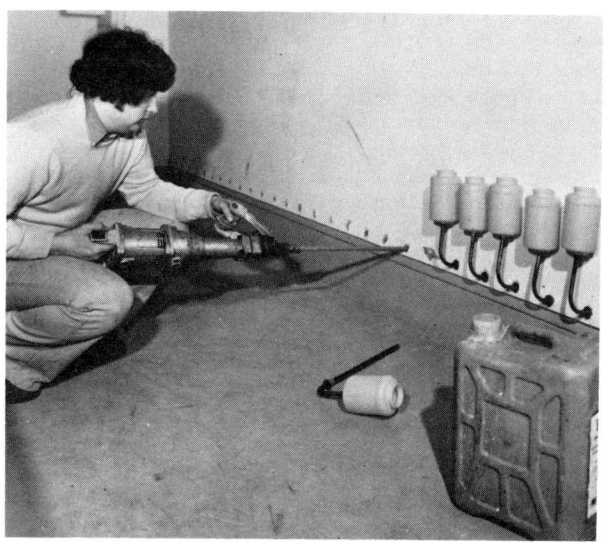

Figure 5.6 Chemical d.p.c. being installed by a Peter Cox operative

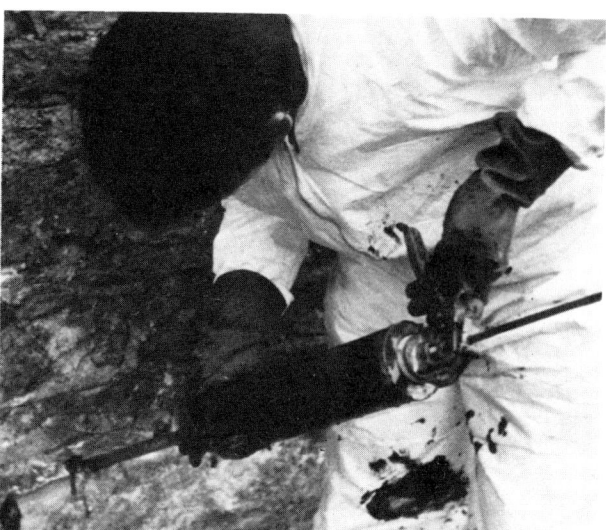

Figure 5.7 Wykmol injection mortar d.p.c. being installed

metals become inoperative after a few years because of electrochemical corrosion. The same could also be true of systems in which an external electrical potential is applied.

Drainage and drying Normally in rubble core walls it is difficult to insert a physical or chemical d.p.c. In these cases, or when for other reasons these treatments are difficult, measures can be taken to reduce the amount of ground water reaching the wall, or to increase the evaporation from the wall. These methods may be used alone, or as a preliminary to lining the wall.

Externally the ground water level can be reduced by placing site drainage near the wall, the area of wall can be exposed below ground or floor level to encourage evaporation, or a moisture barrier can be applied to the outside of the wall. There is a proprietary system designed to increase evaporation by inserting porous tubes into the walls.

Concealing the damp This method does nothing to cure the problem. It merely covers it up; and if there is no danger of structural deterioration as a result, and a relatively short-term life is expected, it is a valid course of action. Either the wall can be drylined with plasterboard on battens (making sure that the battens are preservative treated and that they and the back of the plasterboard are treated to inhibit mould growth), or the plaster can be stripped off the wall and an impervious layer applied before replastering (corrugated pitch or bitumen lathing, or a rubber/tar or bitumen coating – there are a variety of proprietary methods available; see Appendix 1); or the cheapest solution (which is no more than a temporary expedient) is to fix a barrier (pitch paper faced with bitumen or aluminium foil backed by an adhesive) direct to the existing plaster.

All these methods could have the effect of driving the damp higher up the wall. It is therefore recommended that the treatment should extend at least 300–500 mm above the highest defective area of plaster.

There are several plaster treatments that may succeed for a time in preventing dampness in the wall reaching the surface. As has already been pointed out, even when a d.p.c. has been inserted in a wall, the defective area of plaster which may contain hygroscopic salts (plus an area above this) has to be stripped off and the wall replastered. This operation should be delayed for as long as possible to allow the soluble salts to migrate from the masonry into the plaster.

When replastering, it is advisable to use one of the specially developed proprietary plasters recommended for this application (British Gypsum Thistle Renovating plaster). These will not allow the migration of the soluble salts from the damp wall, but provide a quick-drying surface. They also have a resistance to efflorescence and have a hard, durable finish.

Where no new d.p.c. has been installed, replastering in a 1 : 3 cement/sand rendering with a waterproofing additive may provide a short-term barrier to the damp. An alternative treatment is a 1 : 6 cement/sand, plasticised with an air entraining agent, finished with a plaster. This latter mix is likely to be more effective if the wall is composed of soft brick or porous stone. Decorations in all these cases should be porous to limit the rise up the wall of the dampness.

Leaking basements

Leaking basements and their remedial treatments fall largely into two categories: basements which are not built of watertight concrete construction, and those that are.

A. Basements not built of watertight concrete The waterproofing of leaking masonry basements can be an expensive and difficult operation. Often the cause of the leakage is a lack of tanking, or fractured or otherwise ineffective tanking. It can be encouraged by a build-up of water pressure in adjacent waterlogged ground, or it can be a general damp penetration through porous walling or even rising damp in semi-basements (see Rising damp, pp. 34–7). It is essential that the cause of the leakage is established and that there is no danger of confusing condensation for leakage. It is possible that the defect could be linked with a major structural failure, such as settlement, in which case the larger problem has to be tackled before the less critical problem of leakage.

In many cases the dampness will not be impairing the stability or durability of the structure. It is merely causing rapid deterioration of the internal plaster and decoration. The need to embark on a full scale remedial action may be coloured by the ultimate use of the basement. It may even be sufficient to cover up the damp in such a way as to allow a new (and decoratable) wall surface to be built which will provide a permanent lining over the damp wall behind.

Remedial treatments In some cases, particularly where water pressure is evident, an effort should be made to drain the waterlogged ground. Porous drain pipes can be laid round the basement to discharge into a natural watercourse or elsewhere. Even an aggregate-filled trench, adjacent to the basement wall, leading to a soakaway can be effective. However this type of remedial activity is only possible if there is sufficient site area around the building to allow necessary operations to take place without impairing the structure, or that of neighbouring buildings.

Spouting leaks caused by water pressure have been known to have been treated by plugging with a quick setting mix of Portland cement and high alumina cement (1 : 1); but unless the pressure can be reduced, a new leak will invariably start at another point of weakness.

When the basement walls are porous, but not cracked, infusion by a latex-siliconate fluid, either by gravity or under pump pressure can be effective. There are a number of specialist companies in this field (see Appendix 1).

Alternatively the wet walls can be covered up by dry lining the walls with plasterboard fixed on preservative treated battens. Old timber skirtings and the like should be removed from the wall (together with the old plaster), before the battens are fixed. A vapour barrier of polythene sheets should be installed immediately behind the plasterboard and, if possible, the spaces between the battens should be ventilated to the outside air. If quantities of water seeping into the basement warrant it, the cavities can be arranged to drain into a sump. The water collecting here can be pumped away.

The walls can alternatively be covered with bituminous dovetailed lathing (plastered) or waterproof renderings, both after the existing plaster has been removed. In the case of waterproof rendering, this may work on normally constructed walls but there is a danger of curling and cracking of the rendering. A 1 : 3 (Portland cement/sand) mix with a waterproofing admixture can be used, but it should not be applied too wet or be over-trowelled. It should be placed in two thicknesses; total thickness 50 mm. If water pressure is present, the rendering would need to be carried through any abutting internal wall (Fig. 5.8). Special steps may be taken to ensure good adhesion of the rendering in this case.

The installing of tanking, either in the form of 3-coat asphalt or Bituthene or similar membrane, can be undertaken on the inside of the wall, held in place by a blockwork skin. However, this is an expensive method and if there is considerable water pressure behind the wall it may dislodge the blockwork. Asphalt requires a dry background

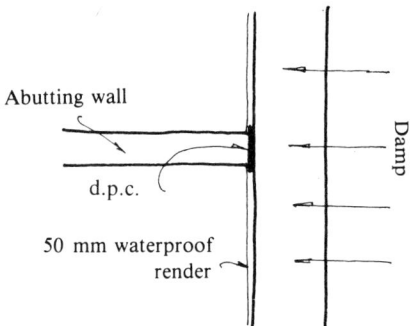

Figure 5.8 Waterproof layer carried through abutting wall

which may be impossible to achieve. This type of treatment also reduces the floor area of the basement. The right place for vertical tanking is on the outside of the wall, but installation of this is usually impractical. As well, it is difficult to seal the bottom of the tanking where it should marry up with the waterproof membrane in the floor.

Basement floors should usually be treated with a d.p. membrane topped with 50 mm of fine concrete. The d.p. membrane should lap with the wall tanking, if any, or be carried 225 mm up the walls.

B. Basements built of watertight concrete These basements are constructed of concrete which does not incorporate a waterproof membrane; the floor is laid without a screed and the walls are not rendered or plastered. Because all parts of the walls are accessible, the positions of the leaks are usually apparent. The floor may present more difficulties. It may help to locate the general area of defect by constructing brick-on-edge bund walls in a grid dividing the floor area into a series of rectangles.

Remedial treatments When the crack or area of honeycomb concrete is located, 45 mm diameter holes are drilled, deep enough to take 20 mm diameter tube sockets, along the line of the crack at 150 mm centres for fine cracks and 600–900 mm centres for wide cracks or honeycomb areas. These holes are then extended at 20 mm diameter to a depth approximately halfway through the wall or floor, or to about 300 mm in concrete over 600 mm thick.

Sockets are then caulked into the holes, being careful not to block the 20 mm diameter hole below. Steel connecting tubes are then inserted into each socket after the hole and the general area of the hole has been cleared of standing water (Fig. 5.9). A liquid detergent is then pumped into the tubes to flush out the cracks and to indicate the more precise outline of the defective area. After this is done, more drilling and the caulking in of more sockets may be necessary. The process continues until the whole area of defective concrete has been established. All detergent is then cleared from the area and pressure grouting can commence through the tubes already used to inject the detergent.

Suitable grouts will vary depending on the particular condition applying in each case. A non-proprietary grout can be used composed of Portland cement (44% by weight), Fulbent 570 (fuller's earth and bentonite – 5%), sodium silicate (1%) and water (50%). Other grout types are proprietary products and should be mixed and used strictly in accordance with the manufacturer's recommendations. These include:

resin grouts (such as epoxy resin)
chemical grouts (such as sodium silicate)
Portland cement and pulverised fuel ash.

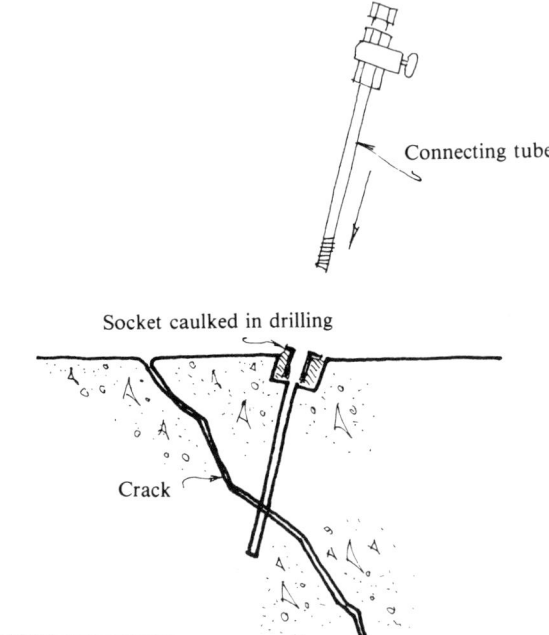

Figure 5.9 Diagram showing the waterproofing of a cracked basement wall or floor

Grout, after straining through a fine sieve, should be pumped into one tube at a time at a pressure of about 1.5 N/mm^2. The cocks on the tubes should be open. When no more grout can be forced into a particular tube, close the cock and continue work at another tube. It is important to return to tubes that have been already grouted, before the grout has set, to clear the tube with a thin steel rod and attempt to inject more grout. This reinjection should be carried on until no further grout can be pumped into the tube.

In the case of very fine cracks, it may be found advisable to leave the detergent in the cracks for about an hour before commencing grouting. Pressure grouting is a skilled job and only companies with suitable experience should be employed. Their methods may vary slightly, but the principle remains unchanged. Professional advice and competent supervision are essential.

Rain penetration of walls

This is another potential cause of damp patches on the inside surfaces of walls. As in the case of rising damp, correct diagnosis is the major problem. Rain penetration may seem to be the cause of damp patches when the real cause is areas of hygroscopic salts in the plaster.

It is also quite easy to mistake condensation for localised rain penetration. Surface condensation is usually associated with poorly insulated walls and often occurs in particular positions in a room where the surfaces are especially cold. Suffice it to say here, that condensation is usually associated with cold, unventilated rooms in buildings which are generally less cold. It is invariably located in corners of rooms or behind large pieces of furniture, where there are pockets of stagnant air. Often the dampness shows up as patches of mould growth. The most easily recognised damp patches caused by condensation are situated in the top corners of rooms, creeping out on to the ceiling. Similar corner patches may be seen at the base of the wall, where they might be mistaken for patches of rising damp. Generally the colder the wall surface, the more likely it is to suffer from surface condensation. Poorly insulated external walls are, therefore, most prone to this defect.

Rain penetration is either due to the general porosity of the walling materials (particularly in solid wall construction) or a defect in the wall or its surrounding accessories which results in greater quantities of water being absorbed by the wall than one would normally expect. Such defects include cracked rendering, or fractured walling material, damaged copings, flashings or rainwater goods, or leaking joints between walling components (between wall panel and wall panel, or between wall and window/door frame). Decayed mortar joints can also result (as all the other defects listed above) in an abnormal wetting of the wall which, in addition to increasing the likelihood of water penetration, can also encourage decay of the walling materials, efflorescence on the wall's external face, washing away of mortar joints and frost damage. It can also endanger any timber built into the wall or corrode iron or steel fixings in the wall. In cavity wall construction a blocked cavity can also cause localised rain penetration. Remedial action in all these cases starts with repairing the defect that is causing the trouble.

General rain penetration of a wall is more common in solid walls than in cavity walls. Penetration due to a porous walling material usually shows up worst on those walls which face the prevailing wet winds (generally in the UK this is the south or west, but the driving rain rose map produced by the BRE – Digest 127 – indicates regional variations in this rule). Walls constructed of hard non-porous bricks can suffer from penetration, usually through the jointing medium. Repointing with a suitable mortar will generally solve this problem; the quality of the mortar being determined by the quality of the brick. Even with a very dense non-absorbent brick it is never recommended that a mix stronger than 1 : 1: 6 is used.

General rain penetration can either be tackled from the outside or the inside. The application of a rendering or some other form of exterior facing material (weatherboarding or tile hanging) can be used, but at the expense of a total transformation in the appearance of the building, which may not be acceptable. Alternatively the walling can be treated with a colourless water repellant solution. These are usually based on silicone resins which line the pores of the walling material and inhibit capillary absorption. Such applications should, however, only be made to walls in good structural condition, since the treatment may increase the likelihood of penetration through cracks or defective joints; walls should also be thoroughly cleaned before treatment to remove all dirt, efflorescence and organic growth; and walls should be as dry as possible, as any water already in the wall will be trapped there by the repellent. If soluble salts are present in the walling, the treatment may induce spalling as the salts are prohibited from moving towards the surface of the wall by the repellent. It is essential, therefore, that all efflorescence is eliminated before treatment.

The selection of a water repellent should be made with BS 3826 in mind, Class A silicone-based materials in this standard should be used on walls of sandstone, all burnt clay products, *in situ* and precast hydraulic cement-based products, cement and cement-lime rendering and asbestos cement; Class B on limestone, calcium silicate bricks and cast stone; and Class C as Class B, but excluding calcium silicate bricks. Class C repellents are useful as a full treatment (or a preparatory treatment before Class A or B) when there is difficulty in drying out the wall.

It should be remembered that water repellent treatments have a limited life (10 years maximum) and will need repeating periodically. BRE Digest 125 gives more information on the subject of water repellents.

The problem of rain penetration can be tackled from inside the building; a course of action often considered more satisfactory. Similar expedients to those suggested in the section on rising damp (page 34) can be employed.

Figure 5.10 Wall joints in Stafford Telephone Exchange have been resealed using Evode silicone sealant

Rain penetration of buildings clad in curtain walling is now becoming a more common problem in refurbishment contracts. Many of the sealants originally used between frames and cladding had a severely limited life and were often incapable of withstanding the natural thermal movement between the different materials making up the curtain walling system. The remedy is invariably to remove the defective sealant, clean and re-prepare the joint surfaces, and apply a correctly-specified sealant whose properties will match the stresses placed on it.

Evode Joint Sealing has undertaken a number of such projects recently including the Rotunda building in Birmingham, in which the joints between the concrete cladding panels and aluminium frames were resealed with Polevo two-part polysulphide sealant, and the Stafford Telephone Exchange, in which 6300 m of silicone sealant was applied to GRP panel and window joints (Fig. 5.10).

Wood restoration and care

The sort of attack to which timber is subjected is particularly insidious because if often works from inside, leaving the outside layer apparently sound. For instance, the slight buckling of the surface of a timber skirting or window frame, or the cracking of its paintwork, may be the first sign of an attack of dry rot (*Merulius lacrymans*); while small perforations in a piece of wood and a tell-tale pile of fine sawdust below, may indicate a live infestation of furniture beetle (*Anobium punctatum*) or the more virulent wood boring insects, the death-watch and longhorn beetles.

Practical problems of refurbishment — and their solution

Table 1 The enemies of wood

Insect or fungus	Timber attacked	Recognition	Remarks
Furniture beetle (*Anobium punctatum*)	Less durable timbers and sapwood	Circular 1.6 mm flight holes	General distribution. Grubs spend 2 years + in the wood
House longhorn beetle (*Hylotruper bajulus*)	Sapwood	Oval flight holes, 10 × 5 mm	Restricted to parts of Surrey and adjacent areas. Grubs 4–7 years in the wood
Death-watch beetle (*Xestobium rufovillosum*)	Often restricted to rotted hardwoods	Circular 3 mm flight holes	General. Grubs spend 3–10 years in the wood
Powder post beetle (*Lyctidae*)	Newly converted sapwood of home-grown hardwoods	Circular 1.6 mm flight holes	General. Grubs are 10 months in the wood
Dry rot (*Merulius lacrymans*)	Softwoods spreading to hardwoods	Cuboidal cracking and cotton wool-like growths	Damp, unventilated conditions
Cellar fungus (wet rot) (*Coniophora cerebella*)	Hardwoods and softwoods	Thin olive-green flat sporaphore	Difficult to detect due to apparently unharmed outer layer. Damp conditions essential
Mine fungus (*Poria vaillantii*)	Hardwoods and softwoods	—	Particularly likely in external joinery in damp conditions
Soft rot fungus (*Ascomycetes and Fungi imperfecti*)	Timber in contact with the ground	—	—

The chief enemies of wood in a building are listed in Table 1 together with the main characteristics of each. Identification of the form of attack, the establishment of the extent of the infestation or infection and its *in situ* treatment is properly a task for the specialist. But the designer of the refurbishment project needs to be on the look-out for timber decay and must beware of importing new, untreated timber into an environment which has already been the scene of rot or insect attack.

There are a great number of proprietary timber preservative treatments which provide protection against both insect and fungal attack. Generally they are based on

Figure 5.11 Fumigation at St. Nicholas, Leicester, to eradicate a death-watch beetle infestation by the Peter Cox method

vacuum/pressure or double vacuum impregnation, using waterborne preservatives based on copper, chromium and arsenic salts or a low viscosity organic solvent. Trade names include Celcure A, Tanolith C, Treatim CCA, Protim Prevac or Hickson's Vac-Vac. Other methods of impregnating timber are less reliable. These include immersion or diffusion.

The treatment of timber *in situ* is a complicated task and one demanding the attention of a specialist contractor. If the work is not performed thoroughly and the attack is not completely eradicated, it will recur after a few months (Fig. 5.11).

Rotted sections of timber or parts that have heavily attacked by woodworm should be cut out and burnt. They should then be replaced by new, preservative-treated wood. It is, however, never sufficient to leave the treatment at that point. The whole adjacent area, whether timber or other material, should be sprayed with fungicide or insecticide.

The only types of preservative that are suitable for application to installed exposed woodwork are those based on tar oils or organic solvents. Waterborne preservatives will tend to leach out due to the washing of the rain, also application can cause the wood to swell. Organic solvent preservatives, on the other hand, can be used on existing external joinery or structural timber without danger of swelling or distortion. What is more, drying is fairly rapid and some preservatives contain water repellents which improve the stability of the wood. Tar oil preservatives have the disadvantage that they create difficulties if further painting or varnishing of the timber is proposed.

There are several proprietary methods on the market for on-site injection treatment of joinery that has experienced limited rotting. First the rotted sections are cut out and replaced, then small ridged plastic inserts are placed in pre-drilled holes (9 mm diameter) in the frame (Figs. 5.12 and 5.13) and organic solvent-based fungicide is injected through them at about 689.5 kPa. Each injector plug has a small non-return valve which maintains the pressure on the injected preservative and assists its dispersion. Inserts are equi-spaced (about 300 mm centres) along the length of all parts of the joinery, or locally collected in areas of particular risk, such as at frame junctions.

After treatment the projecting nipples can be cut off and the holes filled and (after the solvents in the preservative have evaporated) overpainted. The nipples can be left projecting, if future treatment is thought advisable. This method provides an effective way of preventing and arresting decay and can be carried out as part of a maintenance/redecoration programme. It also eradicates insect attack and reduces water penetration of the joinery.

Minor repairs to rotted joinery can be carried out using an external filling compound such as Sylglas. The loose fibrous decayed wood is removed and the affected area

Figure 5.13 Injection of external joinery by the Peter Cox process

filled with a resin-bonded wood flour compound. This usually includes a fungicide which protects the surrounding wood from further rot.

Insect attack is usually treated with a non-staining, oil-based insecticide which creates a toxic layer at the surface of the wood which is said to protect it from further attack for at least 30 years. It also kills the insect larvae as they eat their way towards the wood surface when they are about to fly – the only time in their life cycle when the larvae actually break the surface of the wood.

A recent infestation at York Minster was dealt with by Rentokil using a technique not dissimilar to the fungicide injection method mentioned above. Small tubes were sunk into the mediaeval timberwork through which the insecticide was injected. This ensured maximum penetration of the treatment.

It is very common for infestation and rot to be present at the same time. A typical example of this was in the south aisle roof of St. Mary the Virgin at Woodham Ferrers. Timber beams that had been shaped and chamfered in Tudor times and measured about 300 × 250 mm in cross section were discovered to be badly attacked by insects and fungal decay. Replacement of the beams was out of the question, both from the point of view of cost and the loss of the ancient craftsmanship. As a result it was decided to treat them with Sovereign Deepkill Timber paste. This was spread on any accessible timber face and allowed to work its way into the wood. It is claimed that penetration up to 300 mm is possible and that this is not prohibited by particularly hard and dense timber. Deepkill is also self-skinning, so that the solvents which are vital to the treatment are chemically 'locked' into the timber giving it protection in the future from fungal or insect attack (Fig. 5.14).

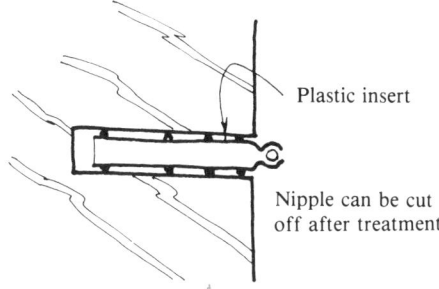

Figure 5.12 Sketch showing the plastic insert injection method

Figure 5.14 Treatment of roof timbers at St. Mary's, Woodham Ferrers, using Deepkill

Timber repair

A method of restoring old timber beams with decay pockets, which avoids the removal of the weakened member and disturbance of the fabric of the building, has been developed by Rickards Timber Treatment. Steel reinforcement is fixed to the underside of the beam, covered by timber mouldings shaped to match the profile of the beam. Epoxy resin is then introduced through cavities in the beam or holes drilled in it for the purpose.

This technique has been applied to a number of historic buildings where the ravages of insects and rot have eaten away at vital parts of the structure. At St. Mary's Church, Winkfield, Surrey, the main support beam of the pyramid roof of the tower spanning 5000 mm was decayed and split. Steel rods were fixed along the underside of the beam and housed in timber shuttering. Epoxy resin was then introduced through the defects in the beam and these were subsequently made good with epoxy mortar (Fig. 5.15).

Another example of this company's ingenuity was in the repairs to the oak roof trusses and ornate carved features of Staple Inn Hall, Holborn, ECI. In this case resin-bonded dowels were inserted and the shakes filled with epoxy resin coloured to match the existing woodwork (Fig. 5.16).

Cause of infection

Finally, one vital point for the refurbisher. It is no good merely curing the infection; its cause must be removed as well. Timber is only subject to fungal attack when its moisture content is for long periods over 20 per cent. Leaks or condensation may have created the right conditions for a fungal attack to commence and it is a fact that insect infestation frequently follows hard on the heels of rot.

The only way you can be sure that rotting timber will not recur in your refurbished building is to remove the original cause of the problem.

Wall renovation and repair

Setting aside any motives of conservation, it is invariably the intrinsic value of the structure of a building that encourages thoughts of refurbishment. In any building older than a hundred years, and very many more a good deal younger, the structure largely consists of loadbearing walls supporting the floors and roof. Hence the walls, it could be argued, are the most vulnerable part of the building; and therefore it is the condition of those walls which will influence the final decision whether or not to refurbish.

However good the state of old walls, it is likely that they will need some renovation, or that their performance may fall short of today's standards in various respects. It is rarely that the walls will fail on grounds of stability, otherwise refurbishment would never have been considered. Except from motives of outright conservation, it can be assumed that the wall is basically structurally adequate, subject to minor structural repair work (see later) or additional strengthening necessitated by the increased load of the refurbished building. The walls, however, will almost invariably need minor repairs, may require cleaning and possibly need additional thermal insulation. They may also require some extra treatment to overcome rain penetration (see previous section).

First the walls should be surveyed to ascertain their true condition.

Wall survey

This examination falls broadly into three categories:

1. Structural.
2. Visual – including the need for elevational restoration repairs and cleaning.
3. Climatic performance – thermal insulation and rain exclusion.

Apart from the normal visual symptoms of structural instability, such as bulges or cracking of the wall, the older masonry wall may suffer from internal cavitation and show no external signs of the defect. As a result, all old masonry walls should be 'sounded' with a hammer to ensure that they are solid throughout. It is particularly old double skin walls with rubble core filling that suffer from internal disintegration. This is the result of long-term rain percolation,

Wall renovation and repair

causing the internal mortar (often of poor quality) to seep out through open joints or accumulate with much of the rubble core as loose fill at the base of the wall. If 'sounding' suggests hollow areas may be present, disintegrated joints should be raked out and probed for voids. It may even be necessary to remove selected face stones and drill deep cores of about 100 mm diameter to assure oneself of the soundness of the wall.

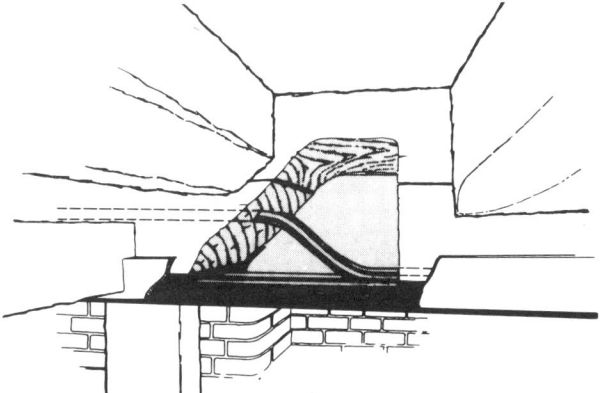

There are some companies which specialise in this type of investigation using some very sophisticated techniques. British Industrial X-rays carry out gamma radiographic investigations and Renofora (UK) ultrasonic testing. Ultrasonic methods are more versatile than X-ray, but require skilled interpretation of the resulting data. The soundings are compared with soundings taken through a section of wall that is known to be solid. Radiography, on the other hand, requires extreme site safety precautions and the local authority will need to be notified that the operation is to take place.

When establishing the structural stability of early cavity walls, it is essential to discover the condition of the wall ties. It is quite likely, particularly if galvanised wire ties were used, they may be suffering from some degree of corrosion.

Structural repair

The repair of voids within the thickness of a wall need not necessarily mean the demolition of that section of the wall

Figure 5.15 Wood repair at St. Mary's, Winkfield, using the Rickards method

Figure 5.16 Wood repair at Staple Inn Hall, Holborn, using the Rickards method

and its rebuilding. The precise action taken will depend very much on the size of the defective area and the general condition of the wall in other respects.

Small voids can be hand grouted, pouring grout into grout cups on the face of the wall and allowing the liquid grout to make its way into the wall interstices.

A more usual hand method is the gravity system. This is particularly appropriate when the condition of the wall is not good enough to allow pressure grouting methods to be employed. The gravity grouting apparatus consists of one or two open pans with base outlets, connected by 38 mm diameter rubber hoses to a 19 mm galvanised iron nozzle with a stopcock. Small holes are drilled into the wall in the vicinity of the voids, roughly 1000 mm apart horizontally and 500 mm apart vertically in a staggered pattern. The wall is then flushed out with water, forced in at a high level

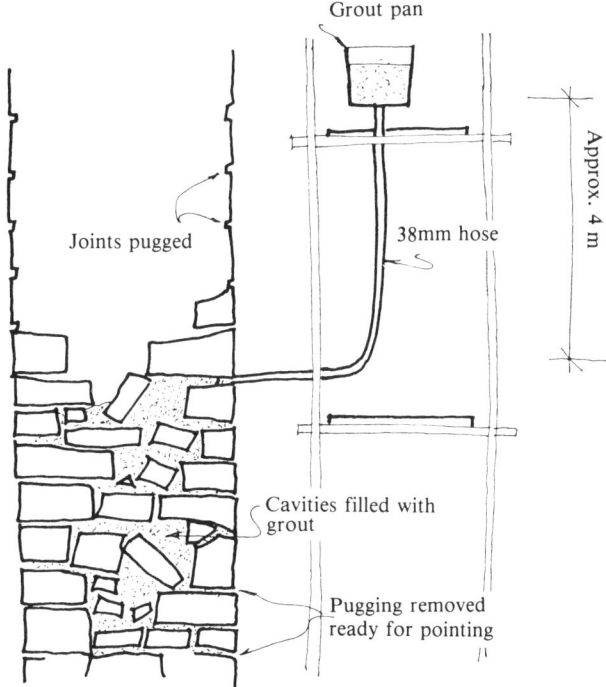

Figure 5.17 Diagram showing gravity grouting apparatus

drilling until it runs out at a lower hole. All leaking joints should then be filled with tow or clay.

The set-up of the equipment should be as illustrated (Fig. 5.17). With the pan placed about 3.5 to 4.5 m above the point of inlet, a pressure of 1.00 to 1.25 kg/cm^2 can be obtained. Airlocks in the equipment should be removed by the use of a rubber cup plunger.

Grout should be allowed to flow into the wall until it begins to seep out of the row of holes immediately above the inlet point. Stop up these holes and continue the operation on the next section of wall. Only 1 m high lifts should be attempted at one time.

An alternative method which can be used on walls in reasonably robust condition is pressure grouting, either by a hand or power operated pump. This method works in a similar manner; but in this case several nozzles are fixed in the wall and the delivery hose is applied first to one, then the next, gradually working up the wall. Power operated equipment produces between 10 and 15 kg/cm^2; whereas hand pumps give considerably lower pressures, which are more appropriate to old masonry walls.

The aerated pressure system (Aerocem) is useful on large scale grouting jobs and is a version of the pumped system using a compressor, mixer, pressure vessel, air lines and delivery hose.

Another method – the vacuum system – involves enclosing the wall area under treatment in an airtight polythene shroud. The air is then evacuated from inside the shroud and the grout in the pan is released through a nozzle sealed to a drilling in the wall and is sucked into the wall thickness. The Balvac system of Balfour Beatty Power Construction is one such vacuum system.

The grout used in these systems can either be of cement or a synthetic resin.

Grouting techniques are useful also to repair cracks in very thick walls, of both masonry and brick construction. Once more pressure grouting should only be used when the wall is generally in good condition and the mortar is sound. The cracks are first raked out and cleaned, then temporarily sealed with clay or weak mortar (leaving a small drainage hole at the base). The crack is then flushed out with water, before grouting commences, tackling 1 m length of crack at a time and feeding the grout into the crack, using a large funnel or similar device. The drainage hole is plugged immediately grout starts to flow out of it. A mixture of 1½ parts of water to ½ of cement and ½ of sand is generally a satisfactory grout mix. After the grout has set the joints and cracks can be raked out and pointed up.

It should be remembered that some cracks in brickwork are best not repaired. Cracks, for convenience, can be divided into three categories:

1. Fine cracks – up to 1.5 mm wide.
2. Medium cracks – from 1.5 to 10 mm wide.
3. Wide cracks – over 10 mm wide.

Wide cracks should always be treated as described above, in wide walls, or by cutting out and rebuilding. The treatment of other cracks will be conditioned by the nature of the brick, strength of the mortar and whether the crack follows the joints or passes through the brick.

Recommended remedial measures can be summarised in this way:

Fine cracks: in porous brickwork, leave untreated;
: in dense brickwork, cut out and rebuild.
Medium cracks: in brickwork built in strong mortar where the crack follows the joint, cut out and rebuild;
: in brickwork built in strong mortar where the crack passes through the bricks, cut out and rebuild;
: in brickwork in soft lime mortar, rake out and fill.

Cracks in internal brickwork should be raked out and filled, except where rebuilding is made necessary in order to stabilize the structure.

Where there is no seasonal movement of the crack, filling is best carried out in a mortar no stronger than the mortar in the rest of the wall. If the crack is subject to seasonal movement (caused by shrinkable clay subsoil or tree roots, etc.) it is preferable to use an oil-based mastic filling rather than mortar.

Cracking in brick vaults in 18th and 19th century cellars is rarely the result of structural failure, unless there has been an increase in heavy traffic passing overhead. Remedial action can usually be restricted to cleaning out and repainting and the removal of any old moisture-retentive finishes.

Cracking caused by differential settlement will have to be remedied by rebuilding. However, this action presents special problems because of the differential downward movement of sections of the wall. This will normally be accompanied by distortion of the surrounding walling and movement inwards or outwards.

The stitching of the crack together with new brickwork will be the normal action required, assuming that further foundation movement is unlikely due to completed settlement, or some remedial action such as underpinning. However, the alignment of the new and old walling may cause a problem. Block bonding can be used on internal walls, but may be unacceptable on some external walls. In these cases a greater section of wall than would normally be disturbed, will need to be rebuilt in order to disguise the irregularities of the repair.

When the wall ties in a cavity wall appear to have corroded, replacement can either be effected by rebuilding one leaf of the cavity wall, or by the use of one of the proprietary double-expansion anchor bolts produced specifically for this operation (Hemax 63 of Harris and Edgar or the Hilti Wall-tie).

If the wall is otherwise in good condition, the filling of the cavity with polyurethane foam insulant will have the effect of adhering the leafs of the cavity wall together and may be considered as an alternative treatment.

Visual repair

The distinction between structural and visual repair is a very fine one. Some disintegration of a building facade may have a structural significance as will, for instance, neglect in the regular cleaning of wall surfaces; but, for the purposes of this book, into the category of visual repair have been collected those defects which primarily affect the appearance of the building.

One of the repairs, which can have structural implications if neglected, is repointing. This is a simple and relatively inexpensive operation when compared with restoration of stonework, but is nevertheless essential if the wall is to be protected from further deterioration. Decaying joints can mean increasing ingress of water into the wall thickness with associated problems of frost damage.

Restoration of stonework, using traditional masonry skills, falls into three broad classifications:

1. Indenting; cutting out the decayed area of stonework and replacing it with pieces of stone from the same source as the original or of a similar geological type with matching porosity, constituents, strength, colour and surface finish. The pieced-in replacement stones should be not less than 75 to 100 mm on bed and, if the backing is likely to be contaminated with salts, the cavity should be treated with sanded bitumen to within 25 mm of the face.
2. Plastic repairs; the rebuilding of the defective area of wall in a mortar composed of lime putty and selected stone aggregates. It may be gauged with white cement or a pozzolanic additive, such as PFA or finely crushed brick dust. Non-ferrous wires and dowels are used as reinforcement and the patched area should not have featheredges.

Figure 5.19 Abbey House, Glasgow; exterior restored using GRC components

3. Resurfacing; dressing back the original stone to a new profile. This is work that can only be tackled by very skilled masons.

There are a number of companies who specialise in repairs to stonework. The Stone Federation will always provide names of suitable companies.

The use of cast stone to replace natural stone elements is a regular feature of stone restoration. In Fig. 5.18 a particularly complex Victorian stair tread, reproduced in cast stone, is being given its final touches in the casting shop of Ian Clayton – a company which specialises in building restoration.

Today there is an increasing use of glass reinforced concrete in stone restoration work. GRC can be cast in moulds taken from the existing stonework to produce perfectly matched elements – and without resorting to the use of highly skilled and difficult-to-obtain stonemasons.

Two recent examples of this type of restoration work have been carried out under the direction of architects, T. M. Miller & Partners of Glasgow. Abbey House, Glasgow, originally designed by Alexander Kirkwood in 1848 in the style of the Italian Renaissance, is a Category B listed building. Over the past 30 years some of the arches on the ground floor elevation have been removed to provide shop windows. As part of a general refurbishment, these arches were reinstated using GRC elements manufactured by Glass Reinforced Concrete of Northwich (Figs. 5.19 and 5.20).

The second example is the refurbishment, after a serious fire in 1978, of the Grosvenor Hotel, Glasgow – part of Grosvenor Terrace, constructed in 1855 and another listed building. The hotel's facade was rebuilt to match exactly its neighbours in the terrace, once more using GRC cast in moulds taken from the stonework in other parts of the terrace.

Figure 5.18 A stone Victorian stair tread being reproduced in cast stone

Figure 5.20 Detail of Abbey House elevation restoration

Figure 5.21 Detail of faience cornice element produced for the Victoria and Albert museum restoration work by Hathenware Ceramics

Similar to the problem of replacing damaged or decayed masonry detailing is that of restoring those materials most beloved of the Victorians – terracotta and faience. The refurbishment of the exterior of such buildings as the Victoria and Albert Museum can be a nightmare unless a manufacturer can be found who still has the specialist skills to make the heavy moulded cornices (Fig. 5.21) and other intricate detailing. One company – Hathenware Ceramics – is such a manufacturer. Among its skilful replacement contracts is the matching up of the Italian terracotta window surrounds in Sutton Place, a manor house constructed in Tudor times.

This company also has the ability to produce any special items in fine earthenware that may be required to match existing work, even down to difficult-to-obtain special sized glazed bricks.

Preservative treatments

Stone decays in the UK mainly due to the crystallisation of salts within the pores of the stone and the washing of rain containing atmospheric pollutants. Both causes of decay are assisted by the porosity of the wall. As a result it seems reasonable that porous old walling, having been cleaned, can be preserved if its porosity can be reduced by using a water repellent. These materials should, however, be used with caution.

Water repellents work by lining the pores of the walling material, thus reducing capillary absorption. They will not, however, solve a damp problem that arises from a different cause than the porosity of the walling. In addition the use of a repellent can prevent a soluble salt build-up within the wall from working towards the surface of the wall and emerging as harmless efflorescence. Incidentally, any efflorescence which exists on the wall should be repeatedly washed, allowed to dry and brushed off, before the wall is treated with repellent.

Early repellents were solutions of emulsions of waxes, oils and resins. Many tended to attract dirt and encourage further decay. They did not penetrate more than a few millimetres into the wall, which made them useless as preservatives.

Today water repellents are based on metallic stearates or silicone resins. The latter are covered by BS 3826:1969.

Three classifications of repellent are stated in this standard:

1. Class A, for use on sandstone, fire clay, cement-based products and asbestos cement.
2. Class B, for use on limestone, calcium silicate bricks and cast stone.
3. Class C, for use on limestone or cast stone – and as a full or preparatory treatment on surfaces which are not dry before treatment. It should never be used on stone with a significant iron content.

Solignum's Impervione and Szerelmey's Stone Liquid No 101 are examples of Class A and B treatments.

A group of materials which preserve masonry by immobilising salts and consolidating friable surfaces are alkoxy silanes which are applied as a low viscosity monomer, and polymerise within the stone to form a polymer network in the out layer of the stone.

Cracks wider than a hair crack and defective mortar joints should be made good before treatment. Also all dirt should be removed and lichens and algae killed with a fungicide (such as Pudlo liquid fungicide or Thaltox Q) and brushed off.

Cleaning

All mass walling materials suffer to some extent from atmospheric attack. Sulpher dioxide in the air, deriving from combustion processes, is absorbed by the rain, forming a weak acid solution. This acid attacks the matrix of many walling materials, particularly natural stone. Marine atmospheres, too, aggravate deterioration. Salt-laden rainwater is absorbed in the walling. On drying out, the salts are left behind, often causing powdering and flaking.

The accumulation of dirt on a wall will act as a reservoir for these aggressive chemicals; and therefore regular cleaning of walls (every three to five years, if possible) is advisable. If a preservation treatment is envisaged, the wall should always be cleaned before any treatment is applied.

Cleaning is basically of three types:

1. Washing and scrubbing.
2. Abrasion.
3. Chemical cleaning.

Washing and scrubbing In this case the grime on the wall is softened by a gentle spray of water, followed by scrubbing with bristle or other non-ferrous brushes. Treatment starts at the top of the building and works downwards. The process is very effective when the dirt is bound to the surface of the wall by water-soluble substances. Today this method is tending to be used more frequently than steam cleaning, because the results of the latter are not appreciably better than those achieved by simple washing.

Generally washing works well on all but heavily soiled sandstone or granite.

Abrasion cleaning When surfaces are deeply soot-encrusted, abrasion cleaning is most successful. Abrasive grit is played, under pressure, on the wall surface, either dry or with the addition of water. Whether wet or dry application is used, the effect is roughly the same. The adhering layer of dirt, is removed along with the immediate surface of the wall, particularly on the arrises. If an over harsh abrasive is used, this can result in blurring of the arrises or the pitting of a soft stone surface. This degree of erosion can be positively harmful and the method is not to be advised for soft stone walling. The creation of dust, too, is a dangerous side-effect, from which the operatives and the public have to be protected, particularly in the case of dry grit blasting. Wet grit blasting does help to some extent to reduce the visible dust. After treatment the walls should be washed down to remove all dust.

Mechanical abrasive with carborundum heads, buffing discs and rotary brushes should only be undertaken by skilled operators. The potential damage that can be inflicted by the unskilled will be costly to repair.

This method is generally rather too aggressive for (and not needed on) moderately or lightly soiled limestone.

Chemical cleaning The choice of chemical cleaner is vital. Many contain soluble salts or react with the walling to form such salts. Only hydrofluoric acid is known not to leave behind soluble salts and it can be used either alone, or in combination with phosphoric acid, which is said to reduce the risk of iron staining on stonework after cleaning. Hydrofluoric acid is a dangerous material, though, and needs careful handling by skilled operatives.

Proprietary stone cleaners should only be used when there is evidence that the cleaner will not have a harmful reaction with the walling material – such cleaners as, for instance, Neolith 625 SS which can be used on sandstone, unpolished granite, brick and terra cotta and has been tested and certified by the Agrément Board.

Limestone and marble are best cleaned by washing, although subsequent brown staining can occur on light-coloured stone. This staining can be reduced by allowing the wall to dry and then rewashing with a water jet. Washing sometimes does not work well on sandstone, granite or slate, although hydrofluoric acid produces a satisfactory result on these stones, as also does grit blasting. Caustic alkali cleaners should never be used.

Concrete, cast stone and rendering respond well to washing, as does brickwork, if the brick is reasonably dense. Chemical cleaners can be used on these materials, provided that they are applied to manufacturer's instructions and that all trace of the chemical is quickly removed after treatment.

Thermal insulation

It is rare that the walls of the building to be refurbished will provide the necessary thermal insulation, either required by the Building Regulations, or dictated by the expense of heating the building adequately.

Two courses of action are obvious. Either one insulates the wall from inside, using a dry lining technique; or one insulates from outside. The former course has the disadvantage of generally reducing floor area by the thickness of the lining and concealing some of the architectural detailing, such as cornices etc.; the latter has the disadvantage of producing a totally changed external appearance. This, of course, need not be a disadvantage at all if the exterior of the building is showing signs of decay, or its

Figure 5.22 Mortimer Hill after Disbotherm System 600 external insulation has been applied

architectural pedigree is undistinguished; it can, however, be a great disadvantage in the case of a listed building. There is, incidentally, a good example of at least one listed building (Mortimer Hill; Fig. 5.22) where external insulation was carried out using the Disbotherm System 600 without remarkably changing its external appearance, but this is the exception rather than the rule.

Internal insulation requires the fixing of a plasterboard/insulation laminated lining to the walls, either on battens, plaster dabs, or on an independent stud framework: Gyproc Thermal board is an example of this type of composite product.

Alternatively ordinary plasterboard can be fixed to battens or studding after an insulation material has been placed between the framing.

The disadvantage of internal insulation is that, because it is on the 'warm' side of the wall, it allows quick internal warm up, but itself has little or no thermal capacity to store heat gained from the internal air. Quick cooling is, therefore, characteristic of this method. The graph of internal temperature is very mobile, compared with the same wall with the insulation placed on the outside. In this case the thermal mass of the wall is slow to heat up, but once warm, will retain its heat and reradiate it to the interior when the internal air temperature falls. A further disadvantage is that unless an adequate vapour barrier is included behind the plasterboard, interstitial condensation can occur, ruining the performance of the insulation.

External insulation methods are becoming very numerous. They fall into three groupings:

1. Those based on an insulating rendering into which pellets of polystyrene have been mixed (Thermorend, Permoglaze; or Thermacote, Blue Circle).
2. Those which involve the fixing of closed cell polystyrene boards on the outside face of the wall and then rendering them to provide mechanical and weather protection (Permoglaze insulation board system; or Styrocote, Cape Insulation Services; or Disbotherm 600).
3. Those which use a quilt of glass fibre fixed to the outside of the wall, covered by a layer of wire mesh reinforcement, and rendered in conventional sand/cement render (Insulath, Tinsley Wire; or Rocksil Thermalath).

They all have the additional advantage of renovating the appearance (and maybe also the rain exclusion performance) of the existing wall (Fig. 5.23). The process is, however, quite labour intensive, because of the intricate treatment needed at the reveals of openings. It is therefore

Figure 5.23 Styrofoam insulation panels being given an impact-resisting render coat

quite expensive. Often, however, it is the only course of action, particularly when the building is occupied or partially occupied during treatment.

Concrete repair

Made in the right way, concrete should last almost indefinitely. This, however, has not always been proved to be a fact in practice, particularly when exposed concrete has not been made in accordance with good practice. A 15-year-old, 5-storey office building in Hounslow – Annabelle House, 28 Staines Road – is a good example of the type of problem that can arise. This building exhibited many areas of bad weathered concrete and blowholes which significantly reduced the cover to the reinforcement. Corrosion of the steel had resulted, followed by extensive spalling of the concrete.

The main reason for spalling is insufficient cover to the reinforcement. The increased loss of alkalinity caused by carbonation of the concrete puts the steel at risk, causing it to rust and build up pressure on the thin concrete cover. Carbonation is a completely normal process and, if there is sufficient cover to the steel, there will be no problem. Carbonation slows up, the deeper it penetrates the concrete.

The very typical defects in Annabelle House were remedied by a systematic method of concrete repair, introduced into this country by Inertol and which has been used by its parent company in Germany for over a decade. Previous cement-based attempts at repair had, incidentally, proved unsuccessful. The process it as follows:

- diagnosis, to determine the type and extent of the defect;
- removal of damaged concrete;
- cleaning the exposed reinforcement to remove all rust (usually by grit blasting);

- protection of the reinforcement (two coats of Icosit Plastic 256 DBP);
- application of a bonding coat (Icoment Additive DBP);
- filling the cavity with repair mortar (Icoment Repair Mortar DBP);
- levelling the repair with a 2 mm coat of Palesit Mortar 520;
- final protective or enhancing coat. Icosit Concrete Cosmetic was used to protect the concrete surface from attack by carbon dioxide, sulphur dioxide and rainwater. This slows down further carbonation. It acts as a barrier to rainwater, while still allowing water vapour to percolate out of the structure. It was applied to existing concrete and repairs alike.

The repair of defective exposed concrete will become a more common refurbishment operation as time goes on. More buildings of the exposed concrete period of the 1960s will reach that critical age at which time earlier shortcomings in the concrete mix or the design of concrete components will be revealed.

Secondary walling elements

This group of building elements comprise all those items which are inserted into a wall to fulfil (as far as the wall is concerned) a secondary function – let daylight in, permit access or ventilation, etc. Of this group, clearly, windows are the most important; firstly, because there are usually more of them than doors or other openings, and secondly, because they have a quite important part to play in the thermal upgrading of an old building so that it can make economic sense, bearing in mind present day standards and costs.

Window upgrading

An existing window may require modification (other than superficial redecoration) during refurbishment because:

- it has been attacked by rot;
- it needs reglazing, either through damage or vandalism, or because the glazing material is not producing the correct standard of thermal or safety performance;
- the window area is simply too large, resulting in high levels of solar gain during the day, speedy heat loss when the ambient temperature is low and glare discomfort for the building's occupants on bright days.

Replacement windows As far as the first reason for modification is concerned, there can be little doubt as to what to do. If rot is extensive, the window needs replacement. This would equally apply to a timber door – but the problems arising in that case are less complex, simply because the number of items involved makes their replacement less financially critical. For instance, the cost of reproducing a considerable number of double-hung sash windows may be prohibitive; but the same elevational appearance can be achieved by a different operational type of window at a fraction of the cost; for example 80 existing sash windows have recently been replaced in an office block in Exeter using pivot-hung windows produced by Yeomans & Partners These match the existing windows in appearance, but give greater resistance to air infiltration (necessary because of the air-conditioning) and were less expensive than reproducing the original windows.

When there is no question of matching the existing windows and general window replacement is envisaged, the choice of window type clearly is influenced by consideration of frame performance – maintenance need, thermal insulation and rain and air exclusion – as well as the frame's ability to carry the type of glazing required. Generally the best long-term economy comes from using the highest quality of windows that the budget will afford. In this respect, plastic windows, because of their low maintenance need, good thermal insulation and generally high insulation performance, should be considered.

Reglazing Apart from reglazing in the same material, due to breakage or sheer longevity of glazing, reglazing is usually undertaken because of the necessity for energy conservation or safety.

Looking at the energy incentive first; buildings with large areas of south-east to south-west glazing will experience a high level of solar gain. This can, in the UK type of climate, be an advantage at some times of the year. In high summer, it can be an embarrassment. Not only can it cause discomfort to the occupants of the building; if an air-conditioning system is installed, it can produce an unacceptably high load on the plant.

Solar transmittance through a single sheet of 4 mm clear glass is 0.86. By contrast, transmittance through a Multiglass Antisun unit (Pilkington Brothers) using two 6 mm sheets of glass can be as little as 0.42, thus rejecting 58 per cent of the solar energy. But reglazing in sophisticated solar control glass or double glazed units can be an exceedingly expensive refurbishment item, particularly if the areas of glazing are large. An alternative could be to apply an insulating film to the glass. This has been found, not only to reduce heat build-up from solar radiation, but also to reduce cold-weather heat loss.

There are a number of solar-control films on the market – such as HAT's Llumar, Klingshield's Sungard and Madico Reflecto-Shield – which can be adhered to the glass without causing optical distortion and which only marginally reduce visibility through the windows from inside the building. Working daylight entering the building (as in the headquarters of Comfort Hotels International in which Scotchtint P19 was used; Fig. 5.24) remains adequate and distressing glare is almost entirely eliminated.

Typical figures for the performance of these films are: solar radiation directly reflected, 46 per cent; absorbed and reradiated, 31 per cent – giving a total of 77 per cent: heat loss due to long-wave infra-red radiation reduced by 37 per cent when the film is applied to single glazing and 23 per cent when applied to double glazing.

This dual thermal function of glazing film has led to one of the 3M films in a modified form (Scotchtint Y-2742) being marketed as a pull-down blind which is said to reduce heat loss through single glazing by as much as 43 per cent when it has a close fit to the frame at the sides and bottom and an air gap between glass and blind of about 6 mm.

Specialist curtaining, also, can control solar input and heat loss. Verosol curtains – the Dutch-manufactured, aluminium-backed polyester net curtaining – can not only reduce the cooling load on the air-conditioning of a building by diminishing solar heat transmission by as much as 60 to 70 per cent, but also have been proved by independent tests to reduce heat loss through the windows by about 28 to 30 per cent. These tests were carried out by the Department of Structural Physics and Heating, Water and Sanitation Technology of the Swedish State Research Laboratory in Borås. Interestingly, these curtains do not impair external vision too greatly (Fig. 5.25).

Double glazing with sealed units is an expensive expedient, but will reduce heat loss and condensation on the inside surface of the glazing. It will, however, do very little to improve the sound transmission through the glazing.

Secondary walling elements

Figure 5.24 Scotchtint P19 used in the Comfort Hotels International headquarters

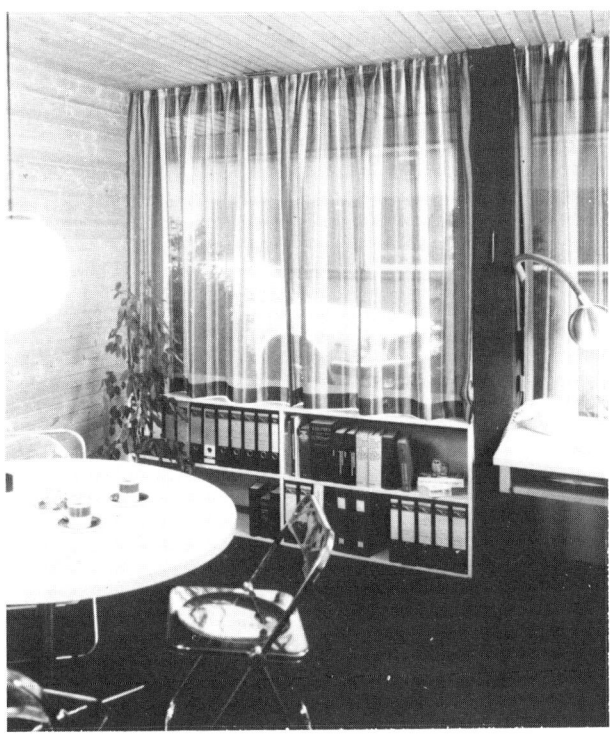

Figure 5.25 Verosol curtains in a modern office

Figure 5.26 Selectaglaze secondary glazing used at Alfa Laval headquarters

A strong case can, however, be made for the use of a secondary form of double glazing when there is an additional need for increased sound insulation. A good example of this is the recent modification of the Brentford Nylons building for Alfa Laval. This building is only 15 m from the M4 and when Alfa Laval acquired the building to house the 300 members of its UK headquarters staff, something had to be done about the noise problem. The Selectaglaze secondary glazing system was chosen (Fig. 5.26).

Because of the larger space between panes of glass in most secondary systems than in sealed double glazing units, the sound insulation is significantly improved. Selectaglaze units used at the Alfa Laval headquarters were tested by the Sound Research Laboratories at Sudbury. From the test data it would seem that an average sound reduction index of around 40 dB can be achieved when the system is used with an average type of primary window and the air space between glazing is between 150 and 200 mm.

Secondary glazing does have the extra advantage of reducing draughts through existing ill-fitting windows – particularly important if a sophisticated air handling system is to be installed.

The other reason for reglazing is safety. Up to 40 000 people in the UK every year suffer from accidents involving glass in buildings. On the face of it, this makes glass a dangerous material, and yet it need not be, if properly used. To ensure safe glazing, the physical characteristics of the glazing materials should be matched with the vulnerability of the positions in which they are used. This design discipline is being increasingly applied to new buildings: it should be equally applied to the design of refurbishment projects, particularly when the use pattern of the building changes in its revitalised form.

Over the past twenty years glass has become a much more significant part of the enclosing envelope of a building. The process started in the early decades of this century and buildings of this period are now prime candidates for refurbishment.

Since the International Modern Movement in Architecture at the beginning of the nineteen hundreds glass has been progressing from a material used in small sizes to let light into the building through a wall constructed of other materials, to a wall cladding material in its own right. But as glass has crept into building, so its potential dangers have been realised.

In any refurbishment project (even when the glazed areas are otherwise unaltered) the properties and position of the existing glazing material should be reassessed in the light of its potential danger to the occupants of the building.

Safe glazing must be the objective; and if its achievement requires the replacement of an area of glazing that appeared otherwise adequate, this should be done.

Annealed glass, because of its brittle nature, fractures upon sudden impact, producing sharp jaggard fragments which can cause deep lacerations. The degree of injury can vary from minor cuts to the severing of major blood vessels, which can lead to death. A small number of fatal accidents from this cause are recorded every year.

The thicker the glass, the stronger and the less likely it is to fracture. Nevertheless all annealed glass – whether sheet, plate, float or figured rolled – can shatter under impact, and therefore if existing annealed glass is in a vulnerable position, it should be replaced during refurbishment.

What then are the vulnerable locations? These could be considered as any position in which glazing is at low level and therefore subject to accidental impact from casual passersby – fully glazed or glass panel doors, glass panels beside doors and similar positions.

There are three types of *safer* glass, and a few types of plastics, which can be used for glazing in these locations:

- tempered glass (or toughened glass); normal annealed glass which has been heated and quickly cooled in a special furnace to produce a glass five times stronger than normal annealed glass. A person walking into a door glazed with toughened glass is unlikely to break it. More importantly, if the glass breaks, the entire pane shatters into many small, and therefore relatively harmless, pieces.
- laminated glass; an annealed glass sandwich with a filling of tough, resilient plastic;
- wired glass; annealed glass into which is embedded a wire mesh which holds the broken glass together (particularly suitable for fire safety purposes);
- clear, patterned or tinted plastic sheets; acrylic sheet has better impact resistance than tempered or laminated glass; polycarbonate can be almost unbreakable. Most glazing plastics are fairly expensive and tend to be less resistant to scratching than glass.
- composite materials composed of plastic sheets reinforced with steel mesh, such as Meshlite from expanded metal.

Safety glazing can be obtained from several manufacturers, such as Pilkington Brothers, Triplex Safety Glass, Alcan Safety Glass, and UBM Glass.

Where reglazing on safety grounds is not practical, the application of a safety film to the glass will improve its shatter resistance. This is a particularly relevant action today with the increasing level of violence.

The lowest grade of Madico/Van Leer safety film, when applied to 4 mm glass, is said to improve its strength to withstand pressures of up to 0.5 bar – strong enough to hold the glazing together in the event of a gas type of explosion.

Draught control The chief cause of unanticipated air infiltration into buildings – and, as a result, heat loss – is through gaps in windows. Windows which are poorly matched to the exposure of the site, or height of the building, can often be the cause of draughts. Worn or poorly fitting windows or doors can also give trouble in older properties. We have already mentioned the example of replacement windows in an Exeter office block, which were necessitated by an air-conditioning system being installed during refurbishment. Clearly the elimination of draughts in this circumstance was essential.

We have also noted that the installation of secondary glazing systems can remarkably reduce the amount of fortuitous ventilation through windows.

If the existing windows are to be retained during refurbishment, it almost invariably pays to include in the contract for fitting the windows (and the doors) with draught excluders – and not cheap strips of foam that rapidly compress and become useless, but more heavy duty devices such as those produced by Schlegel, Duraflex or Kingdom.

The GLC recently commenced a phased contract for the draughtproofing of all the windows in County Hall. This, it is believed, could save as much as 10 per cent on the heating bills. The windows are of an old metal-frame type with hinges and catches that have worn unevenly over the years and the GLC began a research into forms of draught stripping that would take up the differential gapping efficiently.

Finally, Varnamo Rubber draught strips were selected (Fig. 5.27). These are made of EPDM – based synthetic rubber compound (extremely resilient and resistant to environmental degrading) with a closed-cell internal structure and a claimed life of at least 10 years. They come with a strong fibre glass reinforced self-adhesive backing in 100 m rolls.

And while still on the subject of draughts, another recent case study illustrates how the problem of large irregular gaps around existing industrial doors can be tackled. The PSA was concerned at the way the wind whistled through the doors of the nine training hangers at RAF Hereford, Credenhill base.

Cape Insulation Services eventually fitted at 75 mm pile Schlegel draughtseal to the doors. This consists of closely packed polypropylene 'fins' set in an aluminium strip, screwed to the door. The length of the pile provided sufficient versatility to overcome the irregular gapping and it is said that the cost of the work will be recovered in three years.

The problem of industrial doors, which need to be perpetually in use during working hours and which consequently tend to be left open, leading to massive loss of heat, is one which turns up regularly in industrial refurbishment.

The industrial strip curtain door certainly reduces some of the draught and heat loss, while still allowing easy passage of fork-lift trucks and pedestrians. They tend, however, to suffer from swift deterioration due to continual scuffing and the depositing of oil from fork-lift truck masts.

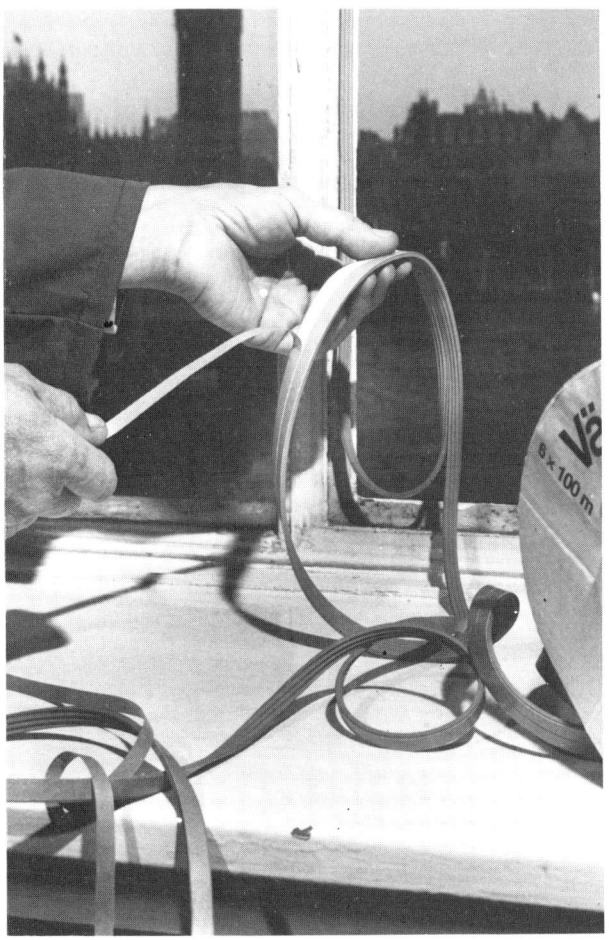

Figure 5.27 Varnamo Rubber draught strips

One of the most effective ways of cutting down window heat loss is the installation of internal insulating shutters (such as the Thermoblind from RMC Panel Products). A 13 mm Thermoblind with a 20 mm air space between it and the 6 mm window glass is said to have a U value as low as 0.9 W/m^2 deg C (as compared with 5.6 W/m^2 deg C for the single glazing or 2.9 W/m^2 deg C for double glazing with a 20 mm air space). What is more, these blinds are easy to install and relatively inexpensive.

Ancient windows renewal

One final aspect of window replacement – the replacement of windows in buildings of architectural merit. Clearly in these cases (and more particularly when the building in question is listed), the replacement windows must match the appearance of the originals.

Some companies specialise in this type of work. The ancient windows of Oxford University's Examination School, consisting of leaded lights set in metal frames, fixed direct into the stonework, were recently replaced by Cotswold Casements with galvanised steel frames. Because the lights had a distinctive rounded shape and consisted of glass now difficult to match in pattern and substance, the work first involved preparing a rubbing of each window in order to record its exact pattern and constitution. The separate panes were then removed and kept in order so that later accurate reassembly was possible.

The replacement steel frames were fixed in the exact position of the originals, fitted to the stonework and finished with a cement seal (Fig. 5.28).

Steel roller or sliding doors provide no suitable alternative, as they tend to be too slow in operation, particularly on busy traffic routes.

A new breed of automatic curtain door (such as that produced by Cape Weiflex) overcomes all the problems. It allows the same speed of entry as a standard curtain door, but does not involve the truck crashing its way through. As the truck approaches the door, a drive motor is triggered off, which parts the curtain just long enough for the vehicle to pass through.

Insulated doors

It should be borne in mind that Part FF of the Building Regulations allows the area of doors to be assumed to be the same U value as the wall in which they are placed. This, however, is so often not the case in practice. In some industrial projects the area of the doors is a significant proportion of the wall area and should, therefore, have as low a U value as is possible.

If, during refurbishment, the doors have to be replaced, it is advisable that the new door has good thermal insulation. A number of manufacturers (notably Crawford Doors and Hillaldam Coburn) market industrial doors with U values at least as low as 0.7 W/m^2 deg C: the joints in these doors are also draught stripped.

Insulating window shutters

A related problem is presented by the over-windowed existing building, with consequent excessive heat loss, but where double glazing or other treatments, mentioned earlier, are inappropriate.

Figure 5.28 Replacement windows in Oxford University's Examination School produced by Cotswold Casements

Similar contracts have been undertaken by this company at the Besse Building and Block D accommodation at St Edmunds Hall.

Moreton & Sons of Winchester is another company specialising in ancient window renewal; its particular skill being in leaded lights. In addition, Tudorglass is producing a form of stained glass, which is a combination of traditional and contemporary skills.

The design of the window is chirographed onto a single sheet of float glass. Lead strips are then bonded to both sides of the pane and the joints soldered by hand. The stain (developed for the company in Germany) is then applied to the sections between the lead work. The window is then hermetically sealed with clear glass to form a double glazed unit with a 12 mm gap between the panes. The stained surface of the glass faces the cavity and is therefore protected from deterioration or damage. A recent example of Tudorglass's work was in the 12th century Norton Priory at Runcorn.

Roof renewal and upgrading

Refurbishment of the roof can comprise an enormous range of activities. On the one hand it can mean the total rebuilding of the roof structure, as well as renewing the roof covering; on the other it can merely consist of the upgrading of an existing roof covering which is generally performing fairly well – the refurbishment being more an improvement of its weathertightness or thermal insulation, rather than a replacement of defective roofing.

Refurbishment, as has been noted before, can often give an opportunity to maintain and improve some parts of a building which are not disturbed by the conversion, but which could, at a relatively small cost, be given an upgraded performance. The roof often falls into this category.

The survey

Once more the survey is all-important. Without a thorough examination of the existing roof, no decision can be made as to the most appropriate form the refurbishment should take and to what extent the roof really needs renewal. A suspected leaking roof membrane can turn out to be a bad case of condensation.

A survey of roof condition should broadly include:

- a detailed assessment of the structure of the roof (are the structural members sound or are they suffering from corrosion, rot or infestation; and will they prove adequate for the future life of the building bearing in mind any new loading, etc. which may be imposed on them?);
- a through examination of the roof covering (is it in good condition or does it let in the water? If it leaks, is it the result of minor local damage – a leaking gutter or a blocked outlet – or is it symptomatic of a major general failure?).
- an assessment of the thermal performance of the roof and the ease with which necessary improvements can be made;
- and finally an examination of the fixings used for the structural members and for the covering elements. A general breakdown of these can invalidate the whole roof. A perfectly good sheeted roof may need to be re-fixed, if the holding-down fixings are showing signs of corrosion.

The extent of flat roof failure is often difficult to establish, while the identification of the *exact* position of a defect is sometimes impossible. On occasions it is expedient to take the very expensive course of stripping off the roofing and replacing it. There is now, however, a technique that can help to amplify a visual examination.

This is based on the phenomenon that all bodies emit electro-magnetic radiation. By making an infra-red scan of the roof, different levels of thermal radiation can be plotted. This identifies areas of poor insulation, probably resulting from wet insulating material. (All insulants operate at full efficiency only when they are dry.) Areas of damp insulation can be the result of a leaking roof membrane, or interstitial condensation. For whichever cause, this scan indicates a roof which is not functioning satisfactorily.

The Thermocore service of Tremco offers this survey technique along with core analysis and determination of bitumen type – all things you need to know before a remedial action can be determined.

Structural renewal

In the case of a major rebuilding of a roof, there are usually compelling reasons for the preservation of the roof form. These could be either aesthetic preference, or simply because the building is listed and any repair to the structure as a whole must not result in an alteration to its appearance.

Sometimes this involves rebuilding the roof framework in more or less its original manner; at other times it may involve the substitution of steel trusses or (as in the case of Pluscarden Abbey, near Elgin, Scotland) the roof trusses can be replaced, once more in timber, but using very up-to-date techniques.

The Abbey was founded in the thirteenth century and was destined for a turbulent life. During the Restoration it passed into secular ownership and only in 1943 was it returned to its intended clerical use. At this time the building was donated to the Abbot of Prinknash, who restored to it a Benedictine way of life, in spite of the fact that the structure was generally in need of large scale restoration.

On the 750th anniversary of the Abbey's foundation an appeal was launched to enable the restoration of the Choir to be undertaken. As a result of this, the steeply pitched (55°) roof was totally reconstructed using Gang-Nail roof trusses. These were fabricated in two parts – a lower section consisting of a truncated truss and an upper section in the form of a fink truss which was to be fixed on top of the truncated truss to give an overall truss height of about 7.5 m.

The two sections of truss were transported to site where they were lapped and bolted together. The composite trusses spanned 10.5 m.

When redesigning the roof structure in this way, it is important to consider the heavy roof covering that the structure may well have to carry. The new, lighter structure may be clad by salvaged or other second-hand traditional roof coverings, such as peg tiles or natural stone slates. This is particularly likely where the external appearance of the roof has to be slavishly preserved.

In addition, extra layers of material, which the original roof did not contain, may have to be included in the rebuilding – sarking felt or boarding, added insulation, etc. Even replaced flat roofs may be subject to additional dead load from extra insulation or protective treatments to avoid ultraviolet degrading of the weathering layer or damage by foot traffic.

One easy way of remedying a leaking roof is to convert it into an inverted roof by laying a new membrane over the existing weathering layer, laying insulation above and loading it with gravel or paving slabs (see later). This however invariably introduces extra weight which the existing structure (unless additionally supported) cannot carry.

In some cases a lighter pitched roof covering can be substituted, provided that its appearance is compatible with the original appearance of the building, or that of its neighbours. Fully-compressed asbestos-cement slates (Eternit or TAC Construction Materials), because of their resemblance of natural slate, can often be used as a substitute for the heavier material. An example of this substitution being recently permitted on a listed building is roof of Wareham Town hall.

Another lightweight alternative to a heavy conventional roof covering is Korrugal Villa Tiles. These are tile-profile units made of aluminium, coated with durable, corrosion-resistant Metallack – an alkyd-melamine stove-on finish. They were developed in Sweden and are marketed in this country by Gränges Essem (UK).

A typical application of this roof covering was at Alexander Court, a three-storey block of 24 flats in Wandsworth. The inability of the roof structures to carry a conventional heavy tile, when reroofing became necessary, coupled with the need for speedy re-roofing with a minimum of disruption, made Korrugal Villa Tiles a wise choice.

Another way of cutting deadweight in roof renewal is to substitute a lighter material for the tradition Code 5 and 6 lead gutters and flashings. In the Wareham Town Hall refurbishment 1½ tons of lead was used in hips, gutters, valleys, flashings and the involved scalloped leadwork of the clock tower. While there was little alternative to lead for the latter feature, in the case of the less exposed and visually important features, a cheaper, lighter substitute could be found.

Nuralite (British Uralite), a bituminous asbestos thermoplastic material, is one possibility; alternatively one of the bitumen felt based flashing strips (with or without a metal facing) can be used. These include Pluvex Ruberflash from Ruberoid Building Products or Sealtite from Marley Waterproofing Products. If a metal alternative is considered wise for reasons of durability and length of life, a zinc/titanium alloy, marketed in the UK under the trade name of Metizinc (Metra Non-ferrous Metals) and which was developed in the United States, could be considered. It can be used to form robust flashings and gutters, or even as a flat roofing or wall cladding, using a conventional standing seam joint between sheets. Metizinc in this form is a worthy alternative for copper or lead; cheaper than either and considerably lighter than lead.

When replacing existing flexible metal roofing, the expensive traditional methods can be adequately substituted by using a factory-produced, fully-supported metal roofing. This is particularly appropriate where a roof surface is visible – on steep pitches or on the bottom slopes of mansard roofs.

Fully-supported metal roofing is soft metal (usually) non-ferrous) factory-laid and jointed over a continuous rigid base. The base is sufficiently strong to accept the fixings of the metal and is capable of itself being fixed directly to the roof framework and spanning between framing elements.

Broderick Structures has over 30 years experience in this prefabricated form of metal roofing. Its Traditional system is particularly appropriate to curved or conical roof shapes, or twisted planes. Its Bonded system is best used on plane surfaces, down to a minimum pitch of 3°. All roofs using Broderick systems and fixed by the company's own operatives carry a full 10-year guarantee.

Another factory-made composite which is ideal for use in exposed locations and which is also produced by Broderick is Lead-Clad-Steel. This consists of 0.75 mm lead cold-roll bonded to 1 mm terne coated steel. It was developed by the Associated Lead Manufacturers and combines the aesthetic and maintenance-free characteristics of lead with the well-appreciated strength of steel. The need for a supporting decking is entirely eliminated; what is more, the traditional threats of theft and vandalism, which are associated with lead-covered roofs, are removed.

A cheaper alternative to these roofings is a metal-faced bitumen felt.

Upgrading performance

A far higher performance is expected of our roofs today then, say, 40 years ago. Then the odd splash of falling water on the factory floor during highly inclement weather would have barely created comment. Today a single drop falling in the sophisticated entrails of a computer could be enough to shut a company down for a month. The thermal performance, too, of the old roof will often be below the acceptable contemporary minimum.

If the existing roof is mechanically sound and generally keeping the water out, it is usually worth investigating the feasibility of leaving the roofing substantially undisturbed and improving its weathertightness and thermal performance by the application of an additional coat or coating.

This is not a sensible course of action where there are inherent faults in the original roof design, such as those which can cause a flat roof membrane to fail (these will be discussed later in this section); or where the roof to be treated contains major defects, such as damaged sheets or slipping slates or tiles on pitched roofs; or old asbestos-cement sheets which have become excessively brittle through age or whose fixings have become corroded.

Before any surface treatment is applied to upgrade a roof, defective areas of roofing must be remedied. A neat method of renovating the fixings of old slate or tile roofs has been developed which involves the adhering of fire-rated polyurethene blocks to the underside of the slates or tiles above the tile lath. This work can be undertaken without stripping off the roofing, assuming that the rear surface of the slating is accessible, and the original appearance of the roof is retained. One of the companies undertaking this type of work is H.B. Roofbond (East Midlands).

There are a number of commercial organisations which specialise in roof surveys and the undertaking of renovation treatment. Each has its own collection of proprietary treatments which can be applied to various types and conditions of roof. Any one of them will precede treatment by carrying out the necessary repairs to the existing roof covering.

Treatments vary from brush- or spray-applied coatings on pitched or flat roofs, to torched-on or loose-lay membranes to flat roofs. Some companies offer a planned maintenance deal along with their roofing service.

A good example of the working of such an arrangement with an industrial company as client is the case of Marshall Engineering of Cambridge. The 50 000 m^2 of roofs have been maintained on a regular and planned basis by Imcco for the last 12 years. Some of the roofs are 40 years old and since the commencement of the Imcco contract all are said to have been trouble-free. One of the treatments which carries a 12-years guarantee when accompanied by a service agreement is the Turnerised Process (Turnerised Roofing Company). This treatment is one of several applied by this company to many roof coverings, both pitched and flat (Fig. 5.29).

First the roof is thoroughly cleaned off and minor repairs carried out. In the case of felt flat roofs this will include cutting and resealing blistering felt and the filling or reinforcing of severely cracked areas and, in the case of sheeted or slated roofs, the refixing of the roofing ele-

Figure 5.29 Turnerised roofing being applied to an asbestos cement roof

ments. On exceptionally porous surfaces a low viscosity black bitumen-in-solvent priming coat is then applied.

A coat of Turnerised solution (a bitumen solution containing fibres/fillers and plasticisers) is then applied to the whole roof in sections. Into this a pre-shrunk cotton membrane is bedded. This is laid with the appropriate amount of 'slack' to accommodate future roof movement. A second coat of Turnerised solution is then laid to impregnate the membrane and complete its bonding to the first coat. Finally, after a curing period, a top coat of solution is applied.

The service agreement, on which the guarantee is conditional, involves the application of a top coat of solution every third year.

This company also has a budget version of this system carrying a 5-year guarantee. Other systems in its repertoire include high performance flat roofing treatments such as the loose-lay PVC membrane systems of Dynamit Nobel – of which more later.

Other bitumen-based brush- or spray-applied coatings have been developed to up-grade leaky roofs. Some of these systems are only recommended for relatively small areas, such as gutters; others can be applied to whole roofs.

The Monoform System from Colas Products has a good track record on larger installations (for instance the 4035 m^2 pitched metal roofs of the potato storage warehouse of Whitworth's Produce of Chatteris). This spray-applied coating, reinforced with glass fibre rovings, was applied to the roof, together with a reflective coating, in less than four days. In this project the junctions between flush GRP rooflights and sheeting had been giving trouble. The metal sheets were showing signs of corrosion in there areas and minor leaking was occuring. First these areas were weatherproofed by bonding with self-adhesive weather-strip, then the whole roof was primed and treated with Monoform over rooflights and sheeting alike. This effectively stopped water flowback at the end laps of the sheets which had led to the corrosion. The whole area was then finished with Decoralt solar control coating, the areas of the rooflights being picked out in brown Decoralt to identify their positions, thus avoiding accidents during future maintenance.

Other companies with similar weatherproofing systems include Robseal and Vel-Va-Lube.

The increasing demand for easy methods of improving the thermal insulation of roofs, without disturbing the operations inside the buildings, has led to the development of a number of multi-purpose spray-on surface treatments which give not only insulation, but improved weathertightness as well. They are all based on polyurethane or polyisocyanurate foams in thicknesses up to 40 mm with a weather resistant finishing coat, which also protects the foam from ultraviolet radiation. One of the major advantages of these treatments is that they add very little dead weight to the roof.

A few of the products in this group are Urecoat (F. Brown & Son), Rubersil (Ruberoid Insulation Services), Roofoam (Robseal) and a similar product from Biltonglow.

The use of a solar reflective paint, such as Solaflect (Screeton Paintmaker), ASP Solar Coat (Vel-Va-Lube) and Blue Circle's Roof Paint, can be used in place of chippings to protect roof coverings in refurbishment and other projects. This makes a significant reduction in the weight of the roofing and perhaps can be used to compensate for some extra weight added during the re-roofing. One square metre of spread chippings weighs about 12.7 kg, whereas Solaflect, for instance, weighs a mere 0.16 kgm^2 (Fig. 5.30).

Figure 5.30 Solaflect reflectant coating at Grays, Essex

Many of these products contain aluminium paste which not only discourages thermal build-up due to solar radiation, but also reduce heat loss by black-body radiation of the external surface at night. Solaflect is a two-coat application and can be applied to old or new asphalt or build-up felt roofs. It has a base of bitumen dioxide which is said to give long life to the treatment.

A method of upgrading both thermal and weatherproofing performance of an industrial roof without disturbing the operations beneath is to install a new sheeted roof above the existing one.

False purlins, either steel or preservative treated timber, are laid parallel to the existing ones, on top of the old roof. New and old purlins are then bolted together through the old sheeting and new sheeting is installed fixed to the false purlins in a conventional way. The void between the new and old cladding is filled with insulation, preferably with a vapour barrier on its underside.

The Robseal Retrofit System and TAC Overclad are two of the commercial systems available.

In any form of roof refurbishment, though, when the stripping off and reinstatement of the old weather proofing layer is not involved, it should be borne in mind that the defects of the original roofing may well reassert themselves. This is particularly true of flat roofs (see below). In over-sheeting as described above, care should be taken to establish that the original roof is structurally adequate (and, of course, will carry the additional weight).

Inherent flat roof defects

In the case of flat roofs, it is essential to establish why the existing roof failed and rectify this cause in the refurbishment. Flat roof membranes usually fail prematurely due to three basic causes:

1. Inadequate drainage.
2. Excessive stress due to structural movement or thermal range.
3. The effect of trapped water vapour below the membrane.

No matter what the weatherproofing layer, all flat roofs should be designed to shed water efficiently. Ponding water on roof surfaces can lead to differential surface temperatures in the membrane, resulting in abnormal stresses. In addition, standing water will seek out even the most minor defects in the roofing that flowing water might miss.

The old flat roof should then be made to drain effectively by adding extra fall pipes or producing falls on top of the existing roof. The method of doing this will depend on the amount of additional weight the roof can carry. A light method is to use composite insulation/plywood panels (such as Aerodeck from Plaschem) on fixed battens. If a single layer roofing (PVC or Hypalon) is specified, the extra weight of the new insulation and roofing, etc. is roughly equivalent to chippings or promenade tiles on the original roof.

The use of a lightweight screed to form additional falls can cause trouble if construction water is trapped below the new roof membrane. A loose-lay roofing, such as Trocal PVC from Dynamit Nobel (UK) or a butyl membrane from Varnamo Rubber, could solve this problem but these membranes need a 50 mm layer of gravel to weight the membrane down and the roof may not to be able to carry this extra load (Fig. 5.31).

If structural movement appears to be the problem, a loose-lay roofing could once more be a solution (subject to the proviso stated above).

Water vapour below the membrane can be caused by interstitial condensation or imprisoned construction water. The latter, once eliminated, will not recur; the former will be a perpetual problem unless an adequate vapour barrier can be introduced into the roof structure on the warm side of the insulation. Invariably the defect arose originally because the roofing membrane itself was the only adequate vapour barrier the roof contained. As the installation of a fully sealed vapour barrier below the roof may be difficult, there are two methods of tackling the defect from above, both of which involve extra weight on the roof.

A respiratory insulation/vapour barrier/first layer roofing composite such as Tekurat (Evode Roofing) can be laid on top of the old roofing, having first punctured the original membrane sufficiently to allow the vapour to escape into the respiration zone of the Tekurat. An extra layer (or layers) of roofing is then applied to the top of the Tekurat.

Alternatively an 'inverted' roof can be formed by repairing the existing membrane (after relieving the pressure of trapped water vapour) topping it with a high performance membrane and laying insulations (a closed-cell polystyrene) on top of the membrane, held in place by a layer of gravel. This is a relatively heavy treatment but one of considerable quality.

At least one of the weathering coatings, developed to protect external polyurethane insulation on roofs from ultraviolet degradation and to provide an EXT F/SAA fire rating to BS 476: Part 3:1958, has been discovered to be an excellent one coat waterproofing membrane in its own right. Applied by brush, roller or airless spray, RAC Rooftex Aluminium Coating from Floorlife-Andek gives reflective, elastomeric membrane with 500–600 per cent elongation for use on asphalt, bitumen felt, slate or asbestos roofing, or direct to concrete or timber decks. This tough aluminish, solvent-based, thixotropic polyurethane membrane can cover hair cracks and can be reinforced with scrim over larger cracks or areas of particular hazard.

Upgrading glazed areas

Existing glazing in roofs usually needs upgrading either because it lacks the required robustness, or because the thermal performance of the original glazing needs improvement.

Figure 5.31 Varnamo Rubber loose-lay roofing membrane after the laying of gravel

Clear polycarbonate is a useful material when high strength, combined with good light penetration, is required. The Belhus Park, South Ockendon, swimming pool has recently been upgraded, during which operation the existing GRP rooflights were replaced by Coxdome CRU rooflights (Williaam Cox). The building had been subjected to considerable vandalism and the new rooflights with their external skin of vandal resistant clear polycarbonate solved one problem, while their inner skin of bronze tinted PVC imparted a warm glow to the pool and the double skin structure reduced previous condensation levels.

Thermal efficiency from roof glazing is difficult to achieve – some would say impossible. However its performance can often be improved. For instance, solar gain can be reduced by reglazing with one of the many solar control glasses and the use of reflective blinds. Heat loss, alternatively, can be tackled by one of the various proprietary insulation glazing composites (Fig. 5.32).

Doulton Glass Insulation's Thermascrene, for instance, consists of two sheets of glass with an acrylic fibre-optic interlayer, the unit being sealed with an epoxy polysulphide or silicone seal. Light transmission is said to be 60 per cent and considerable improvement in heat and noise insulation is claimed.

Thermocell, from a company of the same name, is a transparent 3-ply corrugated roof insulation which has recently been installed in the knitwear factory of Lyle and Scott of Hawick in the Border region of Scotland. A feasibility study carried out by independent engineers confirmed Thermocell's calculations that the reglazing operation would be paid for in reduced fuel consumption in 2 years 10 months. In fact there was an unanticipated bonus. The glazing was in strips over a 16-section, 21-gauge Samco automatic frame knitter, resulting in considerable temperature variations over its huge length. These temperature variations, caused by the differential heat loss between sections of solid and glazed roof, led to frequent breakdown of the machine. Since reglazing, the more even temperature is said to have improved the machine's working to the extent that a saving as high as £7000 per week could result to the company in greater efficiency.

Figure 5.32 Thermocell rooflight insulation

Internal structure remodelling

The internal structure of a building – its floor, ceilings, wall linings and non-loadbearing partitions – is that part of the existing building which, during refurbishment, is most likely to undergo the greatest modification.

It is reasonable to expect that the internal structure, in the new life of the building, will be required to fulfil very different functions from those it performed in its previous life.

Changed usage, and increased or varied occupancy, can bring about requirements for the provision of fire-protected enclosures or fire compartments which did not exist before. The need for varied environmental conditions from one area to another may effect the thermal performance demanded of the internal structure. There may also be a requirement for sound control between rooms – even special security constraints. All these aspects will affect the partitions, doors, and suspended ceilings – their selection and specification, as well as their layout. They may well render the existing partitions, albeit correctly positioned, unsatisfactory for their new role and in need of replacement or upgrading.

In many refurbished office buildings other design decisions will need to be made, such as whether the flexibility of the new occupancy is going to require the use of full height demountable partitions, or whether open office space with low height, sound-absorbent screens is more appropriate.

Partition and ceiling choice

The internal structure of a building not only divides the internal space and provides a means of access and circulation for its occupants and the services which allow those occupants to carry out their tasks in comfort (an item to be dealt with in the next section of this chapter); it also fulfils a number of more discreet functions – such as the control of internal noise levels, the provision of fire-protected escape routes and the facility, in some cases, for a degree of flexibility for the future use of the building.

As has already been mentioned, the function expected of a partition (and the doors it contains) may well change as a result of refurbishment. A change in Purpose Group of the building under the Building Regulations may require an upgrading in the general period of fire protection (for instance, minimum periods of fire resistance are greater in Purpose Group II than III) or the degree of compartmentation within the building. A new pattern of escape may produce requirements for higher fire resistances associated with such features as protected shafts.

Often the various discreet functions which a partition may be expected to perform can appear contradictory. A solid brick or block partition gives good sound attenuation and fire resistance, but is completely inflexible and defies future relocation; on the other hand, it may not always be possible in refurbishment contracts to build what is, in effect, a heavyweight wall on the existing floor structure just where it is required, because of the additional load it will impose on the floor.

If there is no requirement for future flexibility, the timber (or metal) stud partition with plasterboard linings is likely to be the most economic lightweight partition. It can be made to satisfy most fire requirements (except for forming protected shafts of 1-hour or more fire resistance, when the structure has to be non combustible throughout).

Plasterboard gives a Class 0 surface spread of flame, in accordance, with the Building Regulation, and 12.7 mm of Gyproc wallboard on each side gives half-hour fire resistance; two layers on each side (or 19 mm of Gyproc plank) gives a full hour. Table 2 compares fire performance with average sound reduction characteristics of various types of plasterboard faced partitions. It will be seen from this that

Table 2 Fire and sound performance of plasterboard-faced partitions

		1	2	3	4	5	6	7	8	9	10	11	12	13	14	15	16	17	18	19
Structure	*Metal Stud*																			
	48 mm wide	×	×	×	×	×	×	×												
	70 mm wide								×	×	×									
	Timber stud																			
	– 75 × 50 mm																			
	400 mm centres											×								
	600 mm centres												×	×	×	×	×			
	Paramount																			
	50 mm thick																	×		
	57 mm thick																		×	
	63 mm thick																			×
Lining	*Gyproc wallboard*																			
	– 12.7 mm																			
	1 layer both sides	×	×	×	×			×	×		×	×	×			×	×			
	2 layers both sides					×	×			×				×	×			×	×	×
	Gyproc plank – 19 mm																			
	1 layer both sides													×						
Insulation	*Glass-wool blanket*																			
	25 mm thick		×			×			×			×			×					
	60 mm thick			×			×													
	Glass-wool slab																			
	40 mm thick				×															
	Fire resistance (h)	½	½	½	½	1	1	1	½	½	1	½	½	½	1	1	1	½	½	½
	Average sound reduction (dB)	35	41	44	44	43	44	48	35	46	44	30	34	40	35	39	41	29	29	30

generally the less dense partition does not perform remarkably well in respect of sound attenuation.

This is the drawback, too, of many lightweight demountable partitions. While these have the facility for total flexibility, only the high performance partitions can achieve good sound and fire performance – and these tend to be somewhat expensive. It is, therefore, worth assessing just how important future flexibility really is. So often demountable installations are rarely, if ever, demounted. On the other hand many commercial organisations, uncertain of a doubtful economic future, as well as where future technical development will lead them, prefer to invest in future flexibility – and pay the price accordingly.

The selection of a type of demountable partitioning which is capable of receiving a variety of wall finishes is often wise. This allows the same structural system to be used throughout the building, while pointing out the office hierarchy by changed wall panels. For instance the Unilock Firesound partition, used throughout the refurbishment of the 1000 m² City Offices NAFTA in Moorfields High Walk, has suede wallcovering in the reception area – the panels expressing the aluminium uprights – a variety of veneers and textiles elsewhere and – in the director's dining room – a hung panel treatment giving a Regency panelled appearance which conceals the modular basis of the partitioning system.

The before and after photographs of the recent refurbishment at Brixton Estate's head office in Ely Place, London (Fig. 5.33) show the transformation that can be brought about by the skilful use of an up-market partitioning system, a suspended ceiling and a collection of fibrous plaster mouldings, cornices, columns and niches.

The junction between suspended ceiling and partition is critical to several aspects of a partition's performance. If economic demounting is essential, partitions can be fixed between floor and suspended ceiling (assuming the construction of the ceiling gives sufficient rigidity). The suspended ceiling in this arrangement is unbroken, but sound can pass through the ceiling void between rooms. Two ways exist to reduce this; the underside of the structural slab can be treated with sound absorbent material, or a blocking (the full depth of the ceiling void) can be fixed immediately above the partition. When the partition moves, the blocking has to move as well, thus complicating the relocation. It should be borne in mind, though, that Building Regulations may demand the ceiling void be divided down by cavity barriers to prohibit the spread of flames and smoke. The blocking over a partition, in these circumstances, can fulfil both functions. Alternatively the partition can pass right through the suspended ceiling. This method was used at the recent refurbishment of the Midshire Building Society headquaters in Wolverhampton in order to avoid sound bypassing the sound-attenuating Unilock partitions between offices on the executive floor.

Suspended ceilings are rarely required to be demounted (as opposed to modified to gain access through them). Therefore the selection constraints are different from those when a partition is being chosen.

The question of fire resistance of the structure above the ceiling is critical. A suspended ceiling is usually only deemed to add to the fire resistance of the structure above if it is itself non-combustible and if it forms an uninterrupted element, protecting the higher structure. It will, however, create a cavity above it which needs cavity barriers and it will itself require to have a Class 0 surface spread of flame in certain areas. The open grid or screen ceiling (such as those from Formwood) largely avoids the first of these requirements and is appropriate where no additional fire protection is required by the structure above.

Ceilings are either serviced or unserviced. Serviced ceilings are an integral part of the illumination, heating or ven-

Practical problems of refurbishment — and their solution

Figure 5.33 Before and after refurbishment at Brixton Estate's head office, Ely Place, London (Unilock, Project Interiors International)

tilating system, as opposed to merely concealing the works of the services – its ducting, trunking, etc. Serviced ceilings will be dealt with later in this chapter.

Unserviced ceilings can be either jointless (the heaviest method, being generally plaster based), constructed of modular panels or tiles (some or all of which are removable for access to the services behind), or strip or linear panels. Many suspended ceilings today are of the open grid or screen type which do not attempt to form a complete membrane, but merely comprise a grid of slats. A normally horizontal view will give the impression of a solid ceiling; a vertical (or near vertical) view will penetrate the slats to see the objects behind. If the underside of the structure above, and all the service ducts, pipes and conduits are painted black, they become practically invisible and the ceiling gives the illusion of a complete surface into which the lighting has been integrated (Fig. 5.34).

The textile suspended ceiling has recently made its debut on the UK market. Spanoflex from Environaire of Chichester is an economic way of covering up a great deal

Figure 5.34 Formalux 60 ceiling in refurbished office for British Land Property Co development, Kensington High Street

of rubbish above the ceiling. It is a rayon fibre ceiling with a Class 1 flame retardant rating, which is supplied in up to 5 m-wide, pre-moistened rolls, of any length. The damp fabric can be installed quickly and easily; after which it dries, shrinks and stretches to form a smooth ceiling, with no folds, visible joints, seams or air bubbles.

Wall lining

For a considerable time it has been accepted that in shop interior design the walls are treated with linings which, like the partitions, are changed or replaced during refurbishment. Now this habit is creeping into office refurbishment.

So often refurbishment means the wholesale gutting of an interior, pock-marking almost every wall and ceiling with new openings, or blocking up old ones. At the end of the dust and rubble you are left with a series of old plaster surfaces – probably not approaching perfection when the work started, but now almost beyond repair. The temptation to cover the whole thing over with a lining is understandable – if a rather expensive expedient.

In the refurbishment of a suite of offices for British Smelter Constructions in Flyover House, Brentford, Formica post-forming grade laminates were used to produce the 'curved' central column casings and the occasional feature wall lining. This treatment also has the advantage of providing ready-made services distribution cavities around the building, which as we shall see in the next section, is a facility not to be treated with disdain.

Upgraded fire performance

It is worth remembering that when conversion causes an existing element to have a fire rating in excess of its ability, there may be a relatively inexpensive alternative to replacing the element.

The lining of a ceiling with a non-asbestos, non-combustible insulation board (Cape Supalux, or TAC Limpet) can give the required fire resistance to an existing timber floor or partition. Similarly, an existing timber door can be transformed into a fire check door. Assuming the door is a sound flush or panel door, not less than 35 mm thick, it can be made into a half-hour fire check door (30 minutes freedom from collapse and 20 minutes before the passage of flame) by adding one or more sheets of calcium silicate based material. Usually 6 mm of the material screwed to both faces is sufficient, or if protection is required from one side only, one 9 mm sheet fixed on that side, with the panel cavity filled with mineral wool insulation, or plasterboard.

The addition of an intumescent strip, fixed either to the frame rebate or the door edge, will increase the performance to that of a half-hour fire resisting door (30 minutes freedom from collapse and 30 minutes before the passage of flame). In all cases the frame rebate needs to be 25 mm deep, effected by either planting on the lining a new stop, or an additional strip of insulation board on top of the existing stop.

Intumescent shut-offs Intumescent material, such as Palusol Fireboard, can be used to protect the apparent break in the integrity of a compartment floor or wall where a plastic pipe passes through it – a common occurrence in hotel bathrooms and kitchens in high-rise buildings. In the event of fire, the intumescent board expands, crushing the plastic pipe (already softened by the heat) and sealing the hole. The use of strips of intumescent board in outlet louvres to ventilation systems can be used to seal ducts during fire.

The subject of fire protection will be dealt with in more detail in a later section of this chapter.

The serviced interior

There are many influences which conspire to make it an almost invariable fact that during refurbishment the majority of the engineering services will have to be renewed. These influences vary from the simple fact that engineering systems in commercial and industrial buildings appear to have an average effective life of only about 15 years, to obvious new needs of the refurbished building resulting from changed usage or simply greater demands placed on the engineering systems due to altered norms of working conditions or more highly sophisticated communication systems.

Also, because services are dependent upon the use of fuel, most refurbishment programmes today contain a requirement for the installation of systems which will be more economic to run.

Because the design of building services is a very specialised operation, in which each building has to be examined on its own merits, here we will only consider certain aspects of the serviced interior, which are particularly relevant to refurbishment projects, under three broad headings:

1. Restraints imposed by the existing structure.
2. Energy conservation.
3. Package plumbing.

Restraints imposed by the existing structure

Refurbishment is often conditional upon the preservation of the external appearance of the building. This is clearly the case of 'listed buildings', but planning conditions can be imposed on other buildings too. This presents particular problems in the concealment of such excrescences as new lift motor rooms or external plant rooms.

The installation of new lifts during refurbishment is a common problem. Existing lifts invariably need upgrading or even replacing during modernisation. Often, if there is an existing lift shaft, it is too small on plan to house the larger lift car invariably required. The solution to this (and also the protruding lift motor room) can be to install a hydraulic lift. The space in the existing well that was previously occupied by the counterweights of a winding lift is now free to accommodate a larger car, while the lift motor room in a completely new installation need not be constructed popping out of the existing roof, but can be located in the basement, thereby allowing the lift to operate up to its maximum travel capability (usually about 33 m) up to the top floor and with minimal stress to the existing structure.

A recent example of this type of installation is the hydraulic lift installed in the Arazzi Clinic in Devonshire Street, W1. Here a new bed lift replaced an old sub-standard passenger lift; the new lift being housed in the existing well, and the additional size of lift car being made possible by the elimination of the counterweights.

The concealment of new ducting inside the refurbished building can be another major problem, particularly if the ceiling heights of the existing structure are not very generous. With the growing demand for, if not air-conditioning, at least some degree of air handling in present day offices, the need for a more or less continuous service way throughout all floor levels is becoming a normal requirement. Often the level of the window heads prohibits the installation of a general suspended ceiling throughout, as it did at the recent refurbishment of Staple Hall, off Cavendish Street in the City of London. The modernisation of this property involved completely redesigning the building services to provide 100 per cent flexibility of use for the eight floors of open office space. While minor services could be tucked away in skirting trunking, more major services (such as air ducts) proved more difficult to house. Finally modular air-conditioning was installed using 280 Dunham-Bush NC190 fan coil units housed in down-stand sections of suspended ceiling. Hinged access panels were incorporated in the ceiling design for inspection and maintenance.

Where room height and window height will allow the installation of a full suspended ceiling, the servicing advantages can be considerable. The ceiling itself, however, performs a number of other functions:

- it reduces an unnecessarily high ceiling level to one that is more functional today when lower room volumes are permissble due to mechanical ventilation or air-conditioning:
- it hides away unsightly structure and ductwork;
- it can improve the fire resistance of the structure above; and
- it can reduce the noise level in the rooms beneath.

It can even be used to adjust the quality of the acoustics of the room beneath. This was the case of the Hazlitt Theatre at Maidstone, where an old suspended ceiling, which had been obscuring many of the architectural features of the interior, was stripped out during refurbishment and replaced by a lightweight floating canopy. This provided the correct acoustic effect and also revealed once more the decoration on the Corinthian columns and the curving lines of the roof structure.

In many commercial buildings today, the void above the suspended ceiling is often more valuable that the ceiling itself. Not only is it useful to hide cables and ductwork, but often the whole space can be used as a plenum for the air handling system.

The integration of lighting with the suspended ceiling has been a feature of luminaire and suspended ceiling development over the last decade. Such companies as Ascog has produced ranges of flush luminaires that are designed to fit into most suspended tile or panel ceiling layouts. Other companies like F. H. Pride have concentrated on the design of fully integrated ceilings in which the casing or spine of the system/luminaire fulfils a number of functions – carrying lighting and power cables, telephone, video, computer, loudspeaker and clock cables (often in separate enclosures) and in addition extracting air through slots into the ceiling plenum. The movement of air over the lamps allows them to run at their optimum temperature and gives the facility to reclaim the excess heat from the return air. All this lowers the total building heating/refrigeration load. This particular company offers a complete design service to ensure efficient integration of all engineering services at ceiling level.

One of the problems of using the ceiling void as the main servicing route in the building is inaccessibility of island workstations. As open office layouts and complete future flexibility have become the normal requirement of today's office buildings, the difficulties of matching this flexibility with the services (as well as the growing number of services required in the contemporary offices at each workstation) has led to a greater use being made of the raised access floor. This has become particularly necessary with the increasing demand for task ventilation at workstations.

The inclusion of a raised serviced floor is often dimensionally impossible in the refurbishment project; but where the ceiling height permits, it does provide a ready solution to most problems – and provides infinite flexibility for the future. Access floors need not be of great depth. They vary from types like H. H. Robertson Buroplan 300 modular floor, which has a cavity of a mere 65 mm – perfect for cabling, but tight for air handling ducting – to floors with

Figure 5.35 Unilock C5 Datafloor access floor

larger cavities, such as the Unilock C5 Datafloor (Fig. 5.35) which, in its larger form, can become an air-conditioning plenum as well as a service void.

An increased economy can be achieved by the use of a floor-to-ceiling flow in the ventilation and air-conditioning systems. Also with the greater use today of task lighting, there are many commercial buildings in which expensive suspended ceilings can now be replaced by a relatively cheap textile ceiling, such as Spanoflex from Environaire, or a mere visual baffle type of open ceiling such as those produced by Formwood.

Energy conservation

As far as the serviced interior is concerned, much of the energy economy to be achieved is as a result of the skilful selection of the best heating and lighting equipment, the careful design of the various installations and the use of sophisticated control systems to ensure the installations are used economically. These subjects are far beyond the scope of this book; however various general points could be usefully made, bearing in mind that refurbishment provides an opportunity to improve upon the existing engineering installations even if they are undisturbed by the construction work.

Before energy conservation became fashionable – and then an economic necessity – commercial and industrial buildings were designed with little attention to their energy efficiency. Consumption could be in the region of 400 to 500 kW/h per square metre of floor area of prestige office building. With the introduction of energy conservation codes in the States and Europe around 1975, consumption levels were generally fixed at about 200 kW. With an integrated approach to building services design and a building shell engineered for thermal efficiency, levels as low as 120 to 150 kW can be achieved.

The running costs of commercial lighting – particularly in deep plan buildings with a high ratio of floor area to window area – have become a significant element in an office's annual energy consumption – often in the region of 30 per cent.

The traditional method of lighting open office spaces was (and still often is) to flood the area from a series of ceiling luminaires with light in the region of 1000 to 1200 lux. In fact only 700 to 900 lux is needed on the actual work surfaces, while illumination of circulation areas can be reduced to as little as 200 or 300 lux.

To achieve the required illumination levels on work surfaces by the conventional method, the whole area is bathed in light. This gives complete flexibility in the placing of furniture, but tends to produce glare and user eyestrain. Not only is this an expensive method in terms of energy used, but in user satisfaction as well.

Alternative methods based on task lighting, integrated with the furniture and putting the illumination where it is needed – on the work surface – have proved more successful. In these cases low-intensity ambient lighting is provided from a limited number of ceiling luminaires, or indirectly from uplights directed at the ceiling and mounted on columns or walls, or from composite task/ambient fittings attached to the workstations. The latter option usually proves the most economic, with claimed 40 to 50 per cent reduction in running costs and 60 per cent reduction in installation cost. This is, therefore, a method that could well be adopted during refurbishment if there is, within the building, sufficient distribution ability as has been discussed previously in this section.

The correct selection of lamps can have a significant effect on the consumption of electricity. The same, or even better, light output can often be obtained by using a different lamp. In some cases the light output from the alternative lamp is so much greater (while still using less electricity) that a secondary saving can be made by reducing the number of luminaires. Table 3 gives a comparison of energy consumed by tungsten, fluorescent, mercury and high pressure sodium lamps. It also indicates the economies that can be achieved. Some of these lamps are clearly more suited to industrial use.

Because the cost of running lighting is so high today, a number of automatic light switching methods are available. These include operation by photoelectric cell (with delay switches to avoid accidental operation by passing clouds) and automatic time switching. By eliminating human control, which can often be thoughtless and unreliable, savings of from 20 to 40 per cent in terms of energy consumed and payback periods of 18 to 24 months have been claimed for some of the sophisticated micro-processor controls currently available.

A typical system allows programmed switching on a 7-day basis, makes allowances for holidays and can effect fine adjustments such as modifying the switching pattern between workstations near or far from windows.

Similar energy management systems are available to control heating, ventilating and air-conditioning systems. Some units are self-adaptive, modifying the programmed on/off times in accordance with external temperature and adjusting switching to take account of the building's thermal characteristics. They can also 'learn' from their experience and choose more economic switching times. Often they are programmable for a full year, making allowance for holidays and weekends.

Air-conditioning as a necessity of the contemporary office is a concept that is at present being reassessed. Air-conditioning is an extremely expensive facility and may well be unnecessary in many buildings. Openable windows can be prone to misuse by the building's occupants and can be the negation of sophisticated heating controls. It could be argued, therefore, that a ventilation system with fixed windows is a logical and economic approach to the refurbished office; but whether in the UK the need for cooling is proved is open to doubt. It has been said with authority that in the UK air movement is more important than air cooling; and this is often linked to the argument that individuals ideally should have some measure of control over the air movement around their own workstation: one person's healthy ventilation being so often another person's draught. This has led to the introduction of task ventilation/air-conditioning, in which outlets are positioned at workstations with directional and on/off control, very similar to the passenger ventilation in a car.

Heat reclamation is going to become a more essential

Table 3 Comparative energy consumption by various lamps to achieve similar (or better) light output than tungsten lamps

Tungsten lamp (W)	Light output (Lumens)	Alternative lamps:			Light output (Lumens)	Saving including for gear loss (W)
		Fluorescent tube (white) (W)	Mercury* (W)	High pressure* sodium (W)		
100	1,200	40	—	—	2,800	48
150	2,000	40	—	—	2,800	98
300	4,300	65	—	—	4,600	222
300	4,300	—	125	—	5,500	155
500	7,700	125	—	—	8,900	360
500	7,700	—	250	—	13,100	275
500	7,700	—	—	120	8,000	359
750	12,400	—	250	—	13,100	475
750	12,400	—	—	250	22,000	470
1,000	17,300	—	400	—	22,000	565
1,000	17,300	—	—	250	22,000	720

* These lamps are usually used in industrial applications only   (from data supplied by Osram-GEC)

element of commercial and industrial projects, new or refurbished. No longer is heat going to be thoughtlessly exhausted to the outside air by extract ventilation from commercial kitchens and refrigeration equipment, or industrial processes.

The passage of the extracted air through a heat exchanger before finally being exhausted will become a regular feature of design – the heat so gained being used for domestic hot water or pushed back into the general heating system. Similarly, heat fortuitously produced by lighting fittings will also be recycled for internal use.

The heat pump is well worth considering, too as a means of providing usable heat from a low-quality heat source. Quite often a heat pump can make use of the heat reclaimed from exhausted air, improve its quality and reuse it to heat the building or its domestic hot water. Because the heat pump uses less electrical energy than the warmth energy it produces (roughly 1 kW of electrical energy produces over 2.5 kW of warmth), this is a piece of equipment that we shall see used very much more in the future. Operating as a type of reversed air-conditioner, it takes heat from a low-grade heat source (external air, water or earth; or – as above – exhaust air) and apparently produces a miracle by providing the building with more energy than the energy needed to run the equipment. Great development is presently taking place in heat pumps of all sizes from room heaters, similar to a room air-conditioning unit, up to much larger capacity models.

Package plumbing

The advantage of prefabricated plumbing units is that their use allows at least one labour-intensive part of the project to be removed from the site, thus avoiding the delay that is often associated with the installation of engineering services. Also the method ensures a high standard of quality control and a system that has already been tested before delivery to site. There is the further benefit that much of the messy pipework, which can constitute a hygiene hazard as well as an eyesore, is of necessity tucked away inside the plumbing core.

This system can work very well in domestic refurbishment, which is not the subject of this book; but proves more difficult to use in commercial and industrial refurbishment. It is usually only in those building types where there is a quantity of repetitive bathrooms or the like where this type of approach pays off. The obvious examples of this are the hotel and, to a lesser extent, the hospital.

Substantial cost savings have been claimed for the use of pre-plumbed units, such as the Spa-Line system of Schott-Kem, in buildings where a series of bathrooms are to be refurbished. Each unit comprises panels, sanitaryware, pipework and fittings, all fully assembled and tested. Bath, wash basin, shower, w.c., urinal and bidet units can be supplied in protective crates; thus eliminating pilfering and damage. These units are then assembled together to form an integrated bathroom interior. Only the service and drain connections remain to be made. All pipework is concealed behind flush, easily cleaned panels and the company provides a wide choice of design, finish, colour and manufacture of sanitaryware.

About 80 bathrooms at the Embassy House Hotel, Kensington, have recently been refurbished using Schott-Kem units.

The same company also produces pre-plumbed units which are particularly applicable to public washrooms and toilet areas. These units are referred to as the 'K' range (Fig. 5.36) and have similar advantages to Spa-Line – quick installation and clean, hygienic appearance.

Figure 5.36 'K' range pre-plumbed units from Schott-Kem

The requirement for uncluttered interiors with ducted plumbing is being served by several sanitaryware manufacturers who produce fittings designed to fit into or against the surface of plumbing ducts.

Security and safety

A group of elements which, like the mechanical and electrical services, will almost invariably need upgrading during refurbishment, are the security and safety systems. These include intruder alarms, card entry and other access monitoring systems, specialist devices such as security screens associated with cash handling facilities, surveillance devices and fire protection systems.

The majority of these elements are beyond the scope of this book and are clearly the province of the many specialist companies servicing this social need. However it will pay to consider briefly one or two aspects of the fire protection systems and the effect considerations of fire safety can have on refurbishment projects.

In 1979 major fires caused damage estimated at £31.2 m – a rise of more than 150 per cent over the previous year. Any building, new or refurbished, must be designed to reduce the risk of fire occurring; but when, unfortunately, fire breaks out – often as a result of accidental ignition of the contents and nothing to do with the building structure or its finishes – the building should discourage the spread of fire or smoke and be fitted with devices to give the earliest possible warning of the outbreak.

Most accidental fires start by smouldering. The resultant smoke and combustion gases can be as great a threat to life as the fire itself. Not only do these cause asphyxia, they reduce or eliminate visibility on the escape routes, thereby discouraging their use even at times when they would be perfectly safe. Early alarm is therefore vital. Also the escape routes must be protected by structures having adequate fire resistance. The latter aspect will be assured by compliance with the Building Regulations and the advice of the Fire Prevention Officer; the former is a complex subject that is now dealt with in a comprehensive (and up-to-date) manner in BS 5839 (Fire detection and alarm systems in buildings) Part 1:1980 (Code of practice for installation and servicing).

This document, as well as updating the earlier CP 1019, expands on its contents to include:

- a more careful definition of the differing needs of protection of life and the protection of property;
- detailed information on circuit design;
- guidance on zoning of detection and alarm devices;
- recommendations on optical detectors;
- consideration of the substitution of electromagnetic radiation for wiring to transmit signals between components in the system;
- the use of multiplex systems;
- the deprecation of the use of telephones to give fire alarm;
- recommendations for detector and alarm systems in multi-occupancy buildings; and
- initial guidance on self-contained detectors. These devices consist of detector, power supply and alarm sounder.

The document also gives extensive recommendations concerning the selection of detectors. This being one of the most important facets of fire safety, it is worth briefly discussing the various types of detector available.

No one type of detector is equally suitable for all applications. Often a combination of detector types is the best solution. There are three types of detector, each one activated by a different physical change in the environment – heat, smoke or radiation.

Heat detectors are of two types: the 'point' type which is activated by hot gases immediately adjacent to the detector, or the 'line' type which is sensitive along the detector line and reacts to a rise in temperature anywhere along that line. Either type can be designed to activate at a fixed temperature, or in response to a quick temperature rise.

Smoke detectors are either optical, sensing the scattering or absorption of light by smoke, or ionization chamber detectors in which the flow of current through the chamber is interrupted by smoke particles. Sampling and beam-type detectors are effectively 'line' detectors, but 'point' detectors of both optical and ionization types are available.

Radiation detectors are sensitive to ultraviolet or infrared radiation by means of radiation-sensitive cells.

Detectors must be able to distinguish between normal and abnormal conditions within the building and yet be sensitive enough to provide the earliest possible warning of a fire. The rate of response of detectors varies, depending on the type of fire. For instance, a fire which starts with slow smouldering will activate a smoke detector before a heat detector. The reverse would be true in a fire which rapidly evolves heat and little smoke. Flammable liquid fires would normally activate a radiation detector first. Generally a smoke detector seems to be faster to respond than a heat detector, but it is liable to give more false alarms.

Clearly, the likely fire behaviour of the contents of the building needs to be assessed before choosing a type of detector.

Smoke detectors are the general workhorses of fire alarm systems. Ionization chamber detectors respond quickly to clean burning fires, but are more slow to respond to optically dense smoke, such as that generated by PVC fires. In the latter case optical detectors would prove more efficient, but they can give false alarms due to tobacco smoke, which the ionization chamber detector is less likely to do. Clean burning liquid fires, such as alcohol fires, do not activate smoke detectors unless they are fitted with a thermal turbulence device.

Radiation detectors are best suited to this type of fire. They are often used to supplement heat or smoke detectors and are particularly suited to use in high buildings where the device can obtain an unobstructed 'view' of a large space. They are also suitable for surveillance of outdoor storage areas.

Heat detectors are suitable for most types of building, but are not generally recommended for places where large losses could be caused by small fires, as in computer rooms. Fixed temperature types are less suitable where ambient temperatures are low, or may vary over a wide range.

Generally people are the most sensitive fire detectors but, when buildings are unoccupied, reliance has to be placed on the inanimate detector. Today with the considerable number of self-contained units available, there is no excuse for not ensuring that the refurbished building is adequately covered, even without the involvement of considerable lengths of extra cabling.

Fire resistance upgrading

Whenever a structural change is made to an existing building, or its usage is altered, the shell of that building has to be treated to make it comply with the current regulations. This can be an expensive business particularly when the wholesale gutting of the building is not envisaged. It is for this reason that the Fire Research Station has been investigating methods of treating existing structures to upgrade their fire resistant properties while leaving much of their surfaces as intact as possible.

Typical examples of the solutions suggested are shown in the cutaway model illustrated overleaf.

Fig. 5.37 shows a typical 'historic' building situation in which architectural and aesthetic qualities need to be preserved. This is the type of refurbishment contract outlined in Case Study 3 (page 9). The column, for instance, was cast iron (widely used in the 19th century). This might need fire protection without spoiling the architectural effect. Intumescent coatings applied in a minimal thickness have been shown to give structural elements enhanced fire resistance up to periods of one hour. One of these coatings was used on the columns of the refurbished Covent Garden (Case Study 3). The extreme thinness of the coating does not detract from the shapes of the cast iron mouldings (products of this type include Nullifire S30 and Thermo-0 from Fireguard.

In the case of the panelled door in the illustration the weakness is in the panel/style or rail junction. As overlaying with boards, such as non-asbestos Supalux (Cape Boards) or Limpet (TAC), is usually unacceptable in historic buildings, the solution again lies in the use of an intumescent coating.

It was common practice in the 18th and 19th centuries to fit the assemblies into oversize brick openings without sealing the junctions, thus creating continuous cavities behind panelling. These cavities need to be sealed in up-

Figure 5.37 Fire Research Station example 1 (Crown copyright)

Figure 5.38 Fire Research Station example 2 (Crown copyright)

grading work. Cavity barriers may need installing and particular attention must be paid to locations where cavities are continuous through wall or floor construction. A mineral fibre blanket in the cavity would effectively deal with this problem.

In Fig. 5.38 the borrowed light has been reglazed in wired glass, either supported by timber for 30 minutes integrity, or by proprietary methods. The timber bead would need surface protection to prevent ignition by radiation. The example shown uses a noncombustible bead detail similar to the Cape Board's Monolux TRADA fire check channel.

Figure 5.39 shows a fire resistant stud partition in which the facings and infill contribute to the integrity, insulation and stability of the wall. Fibrous infill, derived from silica, rock or blast furnace slag can be used, but its fitting into the wall's cavities is important. The avoidance of gaps is important.

Also illustrated in this example is the cavity barrier above the suspended ceiling, a vital element if fire spread within the cavity is to be avoided. Ductwork may also need upgrading or replacement. The inclusion of a fire damper may be necessary on the line of compartment walls or cavity barriers. These are activated in the event of fire to control the spread of smoke or gas.

An intumescent material, such as Palusol Fireboard, can be used in this type of location. It can be used where plastic pipes pass through compartment floors or walls, A collar made up of strips of this material, wrapped round the pipe where it passes through the structure, will expand in the event of fire, crushing the plastic pipe (already softened by the heat) and sealing the hole. Strips of Fireboard can also be used in the louvres of an air-conditioning/ventilation system to seal off the duct during fire.

Finally in Fig. 5.40 a door assembly is shown which is typical of many that may require upgrading. Assuming the door is a sound flush or panel door, not less than 35 mm thick, it can be transformed into a half-hour fire *check* door (30 minutes freedom from collapse and 20 minutes before the passage of flame) by adding to it one or more sheets of asbestos-free, noncombustible insulation board (Supalux or Limpet). Usually 6 mm of the material screwed to both faces is sufficient, or if protection is required from one side only, one 9 mm sheet fixed on that side, with the panel cavity filled with mineral wool or plasterboard.

The frame rebate needs to be increased to 25 mm deep, effected by either planting on a new stop or an additional strip of insulation board on top of the existing stop. If intumescent strips (2 × 10 mm) are added to the frame rebate or door edge, the performance of the door can be in-

Security and saftey

Figure 5.39 Fire Research Station example 3 (Crown copyright)

Figure 5.40 Fire Research Station example 4 (Crown copyright)

creased to that of a half-hour fire *resisting* door (30 minutes freedom from collapse and 30 minutes before the passage of flame).

One way of upgrading the fire resistance of a timber joist floor without disturbing the existing lath and plaster ceiling below, which may be of architectural merit, is to raise the floorboards and staple sheets of polythene between the joists to protect the ceiling below. Then fix preformed trays of chicken wire or expanded metal between the joists so that there is approximately 15 mm between the plastic sheet and the reinforcement. A 25 mm thick slurry of lightweight aggregate gypsum plaster (metal lathing grade) is then poured over the mesh and the flooring is replaced. This process has the advantage of stabilising the old ceiling which probably has lost its key during the life of the building and would be itself useless as a fire protection for the floor. If the existing ceiling is of no importance, stripping out the old lath and plaster and replacing with new plasterboard would clearly be easier and less expensive; also if the floor needs increasing in fire resistance, a layer of non-asbestos insulation board can be added below the existing ceiling, its fixings also serving to refix the old ceiling.

In all cases upgrading work must be carried out with considerable care and expertise as defects in workmanship or specification can have serious consequences.

One final aspect of fire protection should be remembered. Smoke is one of the major hazards in an outbreak of fire. More people die from asphyxia or poisoning from smoke and toxic gases than from burning. Smoke also hampers escape and the activities of the fire-fighting organisations.

From these points of view the automatic release of smoke and heat on the outbreak of fire is essential, particularly in factory and warehouse buildings. It also helps to inhibit the spread of the fire. Automatic ventilation equipment is therefore likely to be one of those additional safety items that should be installed during refurbishment. An example of this type of inclusion was in the refurbishment of the W. H. Smith store in Worthing (Fig. 5.41). Here the fire officer required architects, John Strong & Partners, to include a 1200 × 750 mm ventilator over the escape stair in the stockroom building. This was wired into the fire/smoke detection/alarm system so that, on the activation of the alarm, the louvres of the Plusaire roof ventilator (Greenwood Airvac) would open and allowed smoke and heat to escape, assisting staff evacuation and fire fighting. It should be noted that these ventilators can often perform the additional duty of providing day-to-day ventilation, opening and closing automatically in response to activation from thermostats, rain sensors or manual controls. In the event of fire all other activation is overridden.

Interior surface refurbishment

By no means all refurbishment contracts involve the repair and decoration of existing plasterwork, such as the magnificently ornate plasterwork of Frank Matcham's Buxton Opera House – part of an extensive refurbishment contract completed in 1979 by Bovis Construction (Fig. 5.42). This is the elite end of a very wide spectrum of interior surface refurbishment.

So often the upgrading contract involves the wholesale gutting of an interior, pock-marking almost every wall and ceiling with new openings, or blocked up old ones. At the end of the dust and rubble there is left a series of old plaster surfaces – themselves probably far from perfect when the work started, but now almost beyond repair. The solution could be the complete replastering of the interior which, apart from the expense involved, introduces a wet, dirty trade into the building programme with consequent drying-out delays. Alternatively the walls can be simply covered up.

At the luxury end of the market, walls can be panelled with a self-finished sheet material. This is a practice that has been common for years in department stores and shops, in which the unfinished internal shell is hidden behind a three-dimensional internal lining which can be simply renewed at will on the changing of occupancy of the premises or as a result of commercial pressure for an updated appearance.

This practice is now becoming more common in office buildings. In the refurbishment of a suite of offices for British Smelter Constructions in Flyover House, Brentford, Formica post-forming grade laminates were used in this way to produce 'curved' column casings and wall linings – not only giving an attractive appearance, but covering up a multitude of sins.

Since the introduction of relatively inexpensive post-forming techniques, the number of joints in laminate linings can be reduced and the unsightly corner condition, where the edge of the laminate is exposed, can be eliminated. In cases such as that in Flyover House, the linings can be made to perform the secondary duty of concealing services which otherwise would have to be chased into the structure.

If lining is out of the question, a poor wall surface can be totally obliterated by the use of a treatment like Tasso's Glass Fiber wallcovering. This Swedish product, readily available in this country, is a white fabric in different textures or weaves. It can be applied rather like wallpaper to virtually any reasonably smooth surface – old defective plaster or even ceramic tiling. After hanging, it is overpainted with gloss, eggshell or flat paint to produce a water and scratch resistant, repairable surface which still carries the pattern of the weave; but is much more hard-wearing than the more normal woodchip paper or many of the soft, present-day wall-coverings. It has a Class 1 spread of flame characteristic to BS 476: Part 7: 1971 and because of its ruggedness has become a firm favourite in hard-wear locations, such as hotel bedrooms and bathrooms. Recently Glass Fiber has been used in a major bedroom refurbishment of the Cavendish Hotel in Jermyn Street, London, and in the creation of the Esselte Headquarters, also in London.

A similar product is Scandatex of Alltek Coatings and there is a wealth of self-coloured natural and synthetic textile wallcoverings from these manufacturers, and others,

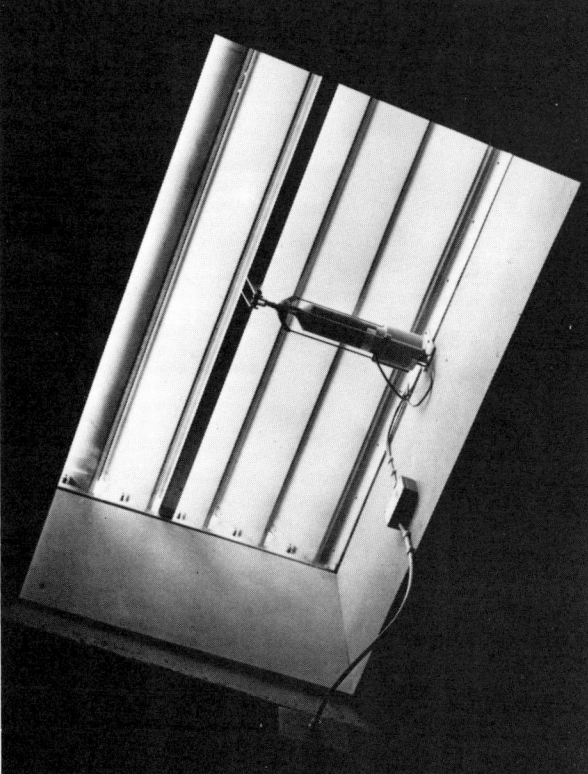

Figure 5.41 Plusaire roof louvre ventilator from Greenwood Airvac used in W. H. Smith refurbishment at Worthing

Figure 5.42 Interior of Frank Matcham's Buxton Opera House

which can cover up slightly defective plasterwork and produce a pleasing (or even opulent) effect, although not giving the toughness of the glass fibre wallcoverings.

A related product, this time originating in the United States, is Flexi-wall plaster-in-a-roll, marketed by Senco Distributors. It is said that this material can be applied to such irregular surfaces as old concrete block walls without filling the joints and it will adequately bridge the irregularities to give a smooth wall surface.

High-build paints are one of the more obvious means of overcoming irregularities in wall surfaces. The number of products of this type on the market has considerably increased in recent years. Not only do these treatments produce a thick textured surface which can obscure the defects of the surface, but they also provide a finished decorative effect, requiring no further paint application.

Some of these treatments do have the disadvantage that they are very difficult to remove when once applied to a surface. What is more, any patching that may result from later alterations is almost impossible to disguise, except by reapplying treatment to the whole surface in question. This has led Berger to recommend that, if removal of its Tartaruga high-building coating is anticipated, a peelable paper underlay should first be applied to the wall. This can be later stripped off, complete with textured finish.

The roughness of texture of these finishes is usually chosen depending either on the effect required, or the degree of damage it is supposed to be concealing – the greater the defect, the more rough the texture.

If it is decided to make good existing plaster surfaces, it might be worth considering a new plaster system from Polycell, Polyskim Trade. The patched area is built up gradually in thin layers of premixed plaster straight from the tub. It can be used without special tools or laboriously acquired craft skills and it can be applied with equal ease to plaster, brick or plasterboard surfaces.

Spray plaster, such as Breplasta, is an alternative method of treating rough surfaces. It can be applied in very thin coats (between 1 and 3 mm) to good fairfaced concrete surfaces. Application is speedy and drying time minimal. Where two-coat work is specified, both coats can be applied in one day and decoration can follow in about 48 hours. Textured surfaces can also be achieved and bond is possible to all types of concrete surface, brickwork and plasterboard.

Finally, two products which may be useful in those damp areas. Respatex Bathroom Panels from Norsk Hydro provide a quick and easy method of lining bathrooms and washrooms. Made up of water-resistant plywood with a facing of high-pressure laminate, these panels have a history of 10 years successful use in Norway. They have a neat tongue-and-groove joint which allows secret fixing to wall battens and there are special internal and external corner fixing extrusions.

If there is a problem with falling ceramic wall tiles in steam atmospheres, it could be that the adhesive used was *water-resistant* rather than *waterproof*. This is precisely what happened at Warley Hospital, Brentwood in the kitchen which produces 3500 meals per day for patients and staff. There were two options; either to strip off the plaster and fix new tiles to the original brickwork with a 'dry set' cement-based mortar – a messy and costly business and one which would have closed the kitchen for a long time; or to remove the old tiles and replace using CeraFix, a Unibond adhesive which is fully waterproof. The latter option was chosen and as a result some 6500 tiles replaced, section by section, with scarcely any dislocation to the kitchen.

Chapter 6

Examples of particular types of refurbishment

Every new refurbishment project is different from the previous one. All have their particular problems, the solution of which is at once the headache and challenge of refurbishment. In this chapter five very different refurbishment projects are briefly examined. Each is interesting in its own right; each represents a particular form of the adaptation of a building to a new updated purpose.

Case Study 5: Theatrical upgrading

Theatre Royal, Nottingham

Client: City of Nottingham
Architects: Renton Howard Wood Levin Partnership
Engineer: Ove Arup and Partners
Main Contractor: Bovis Construction Ltd
Cost: approximately £3.3 million
Date: 1976–78

This project involved the refurbishment and modification of the existing auditorium and stage areas, in addition to the building of substantial sections of new accommodation to contain backstage facilities and front-of-house social spaces.

The modernisation of the Theatre Royal was originally envisaged as being the first phase of a new arts complex, containing in addition a 2200 seat concert hall and a 400 seat multipurpose hall. This proposal was contained in the architects' feasibility study, commissioned in 1975. Due to economic pressures, the major part of the proposals was dropped and only the renovation of the existing theatre and the building of new foyers and backstage areas were proceeded with.

The original theatre which was built in 1865 was designed by C. J. Phipps – one of the most notable theatre architects of his day (Fig. 6.1). However, in 1897 much of the interior was modified by Frank Matcham and the same year the original dressing rooms and backstage accommodation were demolished to make room for the Empire Theatre, constructed alongside the Theatre Royal. A replacement dressing room block was constructed by Matcham which was so appalling that it is said it was one of the chief reasons for the reluctance of touring companies to come to the Theatre Royal. The Empire Theatre, incidentally, was itself demolished in 1969.

The purpose of the project was to construct new foyers (practically non-existent in the original design) and backstage areas, upgrade the stage to facilitate its use in modern theatrical productions, improve the seating (total capacity 1138) and give a proper system of fire exits.

The first part of the work commenced while the theatre was still in operation. This was the building of the new backstage, administration and dressing room block on the site of the recently demolished County Hotel. Minor propping of the structure had to take place in the stage area at this time before the theatre closed. When this finally happened, work could start on parts two and three of the project – the refurbishment of the auditorium and stage. The

Figure 6.1 Portico of Theatre Royal, Nottingham

first job of the contractors was to remove the scenery from the last production!

In the auditorium three private boxes were created at the rear of the dress circle to conceal some obtrusive down-stand beams, the seating and inclination of the upper gallery was improved and the stage boxes and architraves, constructed by Matcham, were stripped out and replaced at a different level by maintaining the sweep of the original balconies. Fibrous plaster decoration was repaired and renovated throughout and new matching sections of plasterwork were installed to the revised balcony fronts (Fig. 6.2). As part of the upgrading of safety standards, new fire doors were installed which perfectly matched the existing doors, while giving 2-hour protection. These were manufactured by Yeoman and Partners (Fig. 6.3). A new ventilation system was also installed.

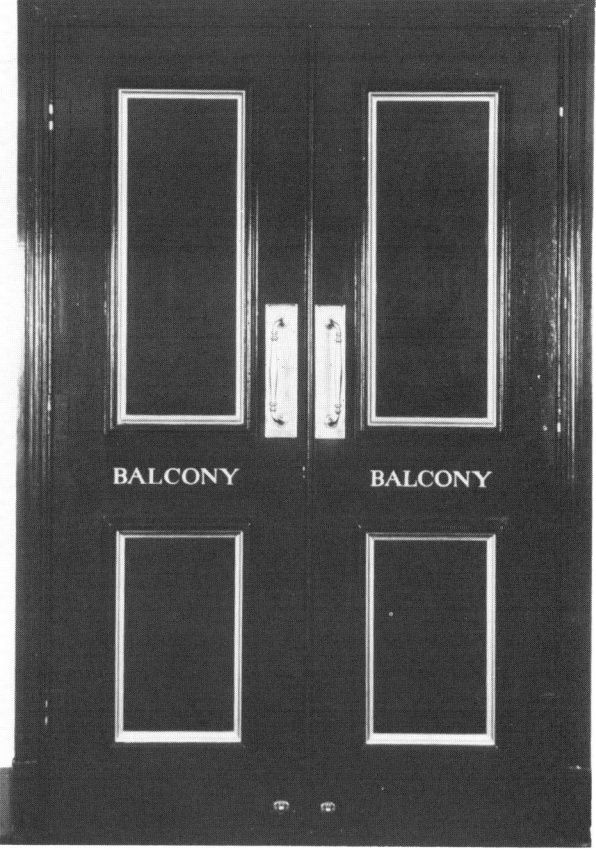

Figure 6.2 New plasterwork, Theatre Royal, Nottingham

Figure 6.3 New fire door, Theatre Royal, Nottingham, manufactured by Yeomans

Examples of particular types of refurbishment

Figure 6.4 New bar area, Theatre Royal, Nottingham

All the old tortuous stairs and bar areas were removed and three large new foyers and bars were created in a new wing to the north of the auditorium (Part 4 of the work) (Fig. 6.4).

In the third part of the work the stage house was provided with an entirely new stage, orchestra pit for full orchestra and wings (which were impossibly small before). A new steel structure was built inside the walls of the old fly tower with new machinery for the flying of scenery. All control for the new lighting and sound systems takes place from the back of the auditorium.

This is an interesting example of a project including new and refurbished elements and in which the parts of the project with innate worth and aesthetic value – the entrance portico, the auditorium and fly tower – were retained; while accommodation which could not be contained within the old shell was planned in new extensions.

Case Study 6: Industrial to cultural change of use

Conversion of a seed warehouse into the Sam Newsom Music Centre, Boston, Lincs.

Architects: Lincolnshire County Architects' Department, Boston Division
Main Contractor: Robert Wallace
Cost: £253,273
Date: 1976–78

A series of unconnected, but lucky, events led to the old Lincoln's Seed Warehouse on the river frontage at Boston, Lincolnshire, being acquired by the Lincolnshire County Council and converted into the Sam Newsom Music Centre. The existing Boston music centre, which formed the base for the county's advisory and instrumental teaching staff, also housed the Boston College of Further Education's

Figure 6.5 Lincoln's Seed Warehouse, Boston (courtesy The Architectural Press Ltd and Lincolnshire County Council (building owner); architect, Lincolnshire County Architect's Department)

Figure 6.6 Recital Hall, Sam Newsom Music Centre, Boston (courtesy The Architectural Press Ltd; photographer Richard Bryant)

Music Department and acted as a focus for the area's musical activities. It was, however, about to lose its existing accommodation due to a road development. The warehouse, which is a Second Schedule listed building and occupies a central position in the town, fortuitously came on the market. The initial feasibility study established that it could be converted to house the accommodation for a new music centre as scheduled by the County Education Authority.

The warehouse had been built in three phases, in 1741, 1780 and 1810, of simple post and beam construction within a substantial brick shell. It had a double collar pantile roof and was typical of much east coast vernacular architecture of that time (Fig. 6.5).

Although considerable settlement had occurred, causing distortion of many floors, it was decided that this had been caused mainly by the third phase of the building – the addition of a drying room on the roof, which had overloaded the foundations, particularly of one of the cross walls. This drying room was to be demolished during the project, together with the failed cross wall, thus making room for the recital hall which formed the focus of the refurbished building (Fig. 6.6). The whole structure was strengthened at high level by casting a continuous *in situ* ring beam to support a new steel framed roof, at intermediate level by a concrete diaphragm floor to the recital hall and at ground floor level by casting an *in situ* concrete retaining wall to the riverside wall. Internally the timber structure was, as far as possible, retained and given a 1-hour fire resistance by means of a white intumescent coating.

Existing internal brickwork was sandblasted to remove the original limewash and new partitions were built of common brickwork with an emulsion paint finish.

Externally the brickwork was cleaned by low pressure water washing and was repointed. Any exterior repair to the walls was carried out in the original brick, or using bricks of a similar age or type. Replacement windows were designed to match the existing ones and secondary glazing was added to improve sound insulation. The roof was renewed in pantiles to match the old tiles.

A ventilation plant was installed to replace natural ventilation eliminated by the need to reduce external noise penetration.

This is an imaginative reuse of a building whose chances of survival in a usable, well-repaired form were sparse. As it is, this old warehouse has made a music centre of considerable charm with many similarities to the Maltings Concert Hall refurbishment at Snape (Fig. 6.7).

Case Study 7: Refurbishment with occupants *in situ*

The Royal Liver Building, Liverpool

Client: Royal Liver Friendly Society
Designers: Arup Associates
Management Contractors: Bovis Construction Ltd
Cost: approximately £5 million
Date: 1977 to 1980

The problem of upgrading and adaptation of buildings while their occupants are in residence was illustrated dramatically in the refurbishment of the Royal Liver Building at Pier Head, Liverpool.

One of the most notable landmarks on the Liverpool waterfront, this granite-faced building boasts an impressive twin-towered outline and clockfaces larger than those of Big Ben at Westminster (Fig. 6.8).

The work, undertaken by Bovis Construction under a management contract, consisted of the complete modernisation of the interior of the building, one floor at a time, while the tenants remained undisturbed on the other floors. This meant considerable problems for the contractor, who had to maintain the building's services at all times to all floors, except that on which work was actually taking place. Other problems included 1500 m^3 of demolition rubbish which had to be removed from each floor in spite of very restricted access; also the client demanded that the external appearance of the building should give as little evidence as possible of the work progressing inside.

Figure 6.8 Royal Liver Building

Figure 6.7 Snape Maltings after refurbishment

All eight above-ground floors of offices were upgraded, together with additional work to accommodation in the famous towers. New suspended flooring was fitted, curtain walling was installed to two large internal light wells, plumbing installations and engineering services were completely replaced and refurbished, and about 1700 replacement metal windows were fixed.

Case Study 8: Conversion from domestic to commercial use

Nos 31 and 32 Curzon Street, London W1

Client: Pilkington Brothers Superannuation Fund
Client's agent: Herring Son and Daw
Contractors: Yeoman and Partners Ltd
Date: 1980

Two historic Mayfair mansions were converted and refurbished to form 560 m² of distinctive office accommodation. The properties – Nos 31 and 32 Curzon Street – are situated on the south side of this historic thoroughfare, once the site of London's old May Fair. The refurbishment has ensured that this part of the street frontage, at least, will be preserved (Fig. 6.9).

The properties concerned once had owners of considerable distinction. Rufus Isaacs (1860–1935), first Marquis of Reading, Viceroy of India, Lord Chief Justice and President of ICI, used to occupy No. 31; while next door at No. 32 lived the Howe family – the third Viscount Howe (George Augustus Curzon) who lived there in the 18th century lent his name to Curzon Street. The properties had fallen into disrepair and multi-occupancy in recent years.

When the Pilkington Brothers Superannuation Fund acquired the two properties in 1979, it was decided (in consultation with the planners and historic building authorities) that the buildings should be completely restored. Behind the facades, the rooms on the two lower floors had been

Figure 6.10 31–32 Curzon Street; interior

Figure 6.9 31–32 Curzon Street; exterior

Figure 6.11 31–32 Curzon Street; garden elevation

clumsily adapted to the changed occupancy and were generally suffering from damage and lack of maintenance.

The contract for undertaking the work was awarded by Pilkington's agents to Yeoman and Partners, who have tended to specialise in this type of prestige refurbishment. The work on site was carried out in about 6 months; 2 months less than the programmed contract period.

The main reception rooms were restored to their original proportions (Fig. 6.10) while providing the high standard of supporting amenities that would be expected by a tenant prepared to pay a top West End rental.

In places the existing joists had to be levelled and new floor boarding installed. The building was completely re-wired, the central heating system was upgraded, new sanitaryware was fixed and the lift system updated with a new 4-person passenger car.

Probably the most interesting part of the scheme was the restoration work that was carried out to such features as the pinewood box shutters and sliding sash windows and the recreation of the interior decor with all its intricate plaster detailing. Many of the existing cornices, dados and fireplaces had been damaged or disturbed during the previous adaptation of the properties. Yeoman's team of skilled plasterers repaired and matched up these details, precisely recapturing the spirit of the two mansions when they were in their heyday (Fig. 6.11).

Case Study 9: Disused factory to training centre

Glass insulators factory conversion

Client: Pilkington Brothers Ltd ·
Architect: Building Design Partnership, Preston, in association with Pilkington Central Engineering Department
Engineer: A. C. Robinson, St. Helens
Builder: Holland, Hannen & Cubitts (Northern) Ltd
Cost: approx. £1 million
Date: 1972 to 1979

In 1971, when the manufacture of high voltage glass insulators ceased, Pilkington Brothers was left with a redundant (and far from attractive) building in Alexander Drive, off Prescot Road, St. Helens (Fig. 6.12). This old production building was located close to the head office and was an ideal location for the Group Training Centre which the company had wanted to establish in one centralised position.

It was decided to house all the activities of Group training and development on this site, giving precedence to the establishment of the Engineering Training School for apprentices. This was to form the first phase of the development and, because of the weight of the machines used in this school, it seemed logical to locate it on the ground floor of the refurbished production building.

The existing steel structure had been designed originally merely to support the roof and the external wall cladding. No intermediate floors had been envisaged and

Figure 6.12 Glass insulator production building, Pilkington Bros

Figure 6.13 New training centre, Pilkington Bros

therefore the additional first floor, that was proposed in the conversion, had to be erected inside the original structure and resting upon its own independent frames, whose columns were supported on bored pile foundations. This work was undertaken in the first phase of the refurbishment so that later work could proceed without interfering with the training programme in the Engineering Centre on the ground floor. Phase 1 of the conversion was completed in 1973.

A year later the second phase of the project was authorised, but for economic reasons the work was suspended in 1975 just after a new roof had been constructed below the original roof and before the structure above the new roof was demolished.

In 1977 Pilkington Brothers' Board authorised a new start and the transformation of the ugly production building was completed. The existing roof and all steelwork above the new flat roof was demolished and a totally new elevation was created, using Pilkington products. Reflectafloat reflective solar control glass, formed in double glazed units, was used to clad three elevations at first floor level, below a strip of GRC cladding panels, constructed by Glass Reinforced Concrete of Northwich using Pilkington Cemfil alkali resistant fibre (Fig. 6.13). The rest of the glazing of the building was in single skin panes separated by a 100 mm air gap from black neoprene faced 75 mm Fibreglass Factoryliner. High levels of insulation were obtained from the 80 mm insulating core in the GRC panels and the lining system.

Entrance to the Engineering Centre is at ground level; the upper floor is entered from an existing bridge leading to the company's head office complex. Internally, the air-conditioning and lighting systems incorporated heat recovery. This is particularly important in the new colour television studio where a maximum of 30 kW of lighting is sometimes in use. This studio is isolated as far as possible from external noise and a separate air-conditioning system serves this room.

This example shows how use can be made of a basic structure to create an entirely new building for a totally different use and with a dramatically changed appearance.

Chapter 7

The handover

Any building owner or tenant occupying a building for the first time should ideally be provided with two pieces of information: a maintenance manual which lists the finishes and equipment in his building and explains their maintenance and care in use; and a full set of as-built drawings. The latter is particularly important in the case of the refurbished building, where special structural methods may have been employed during the conversion and where the methods and routing of services may not be immediately obvious. In effect this dossier is similar to an up-market handbook which tells the new car owner about his car and how to look after it. It will not solve all his problems, but it will at least solve some of them; and point him in the right direction in other cases.

The maintenance manual prepared by the architects, Abbey and Hanson Rowe and Partners, for the Lloyds Bank building in Leeds (Case Study 2) is a good example of a document of this type.

Maintenance manual: Lloyds Bank Building, 31–32 Park Row, Leeds

Part 1 of the manual gives a brief description of the building and details of the refurbishment contract, including names, addresses and telephone numbers of the consultants involved, local authority approval details and who were the statutory undertakers. This part ends with a comprehensive list of subcontractors and suppliers used on the contract, together with names and telephone numbers of persons to contact in an emergency; for instance in the event of failure of the lift installations etc. There is also a section to be filled in by the Landlord (Lloyds Bank) of persons who should be informed of failures in the building or its installations by tenants of the building.

Part 2 concerns itself with the finishes of the building. It lists all the finishes used and gives details of the subcontractor or supplier involved. It also sets down an area-by-area schedule of where these finishes occur and gives basic details concerning the structure of suspended ceilings and plumbing ducts. It then goes on to furnish regular cleaning and maintenance instructions of these surfaces in considerable detail. The first three parts of this maintenance section are here quoted in full as an example of the type of information that a building occupier would find extremely useful. These parts deal with cleaning and maintenance of the exterior, and similar advice concerning the cleaning and maintenance of the stainless steel elements internally and floors. The details included concerning door furniture should also be noted.

10.01	*Externally*
External glazing	*Only cleaning should be necessary. An annual inspection should be carried out to ensure that the mastic between the glass and the frame is intact, check that the mastic has not been gouged out allowing water to load or penetrate the glazing on the bottom edge of the windows.*
Granite	*The polished granite cladding is virtually maintenance free and only requires*

The handover

washing with luke warm water and mild detergent then rinsing down with clean water to remove surface deposits of grime and grit, then dry thoroughly.

We recommend that the external granite be washed, (by the Window Cleaning Contractor) annually.

No abrasive powders or polishes should be used otherwise the highly polished surface will become dull. To remove graffiti the following treatments may be used:

(a) *Cellulose or other spray paints – Carngel*
(b) *Felt pens or spray paints – Nitromors*
(c) *Felt pens – Methylene chloride*
(d) *Crayon or felt pens – Acetone/isopropyl alcohol mixture*

10.02 Internally

Stainless steel signboard and other stainless surfaces

The following recommendations are made by The Stainless Steel Development Association:

1. Removal of dirt
(a) Soap or detergent and water, apply with a sponge, rinse with clean water, wipe dry if necessary. Household detergents such as Tide or Omo can be used.
(b) Mild abrasive cleaners can be used for a brush or dull polished finish, the abrasive should be rubbed in the same direction as the polish lines. Rinse with clean water and wipe dry. Abrasive will damage highly polished finishes.

The abrasive cleaners can be pumice or scouring powders such as Ajax, Vim etc. Some proprietary household powders contain chloride bleaches which if left on the surface might cause pitting. Thorough rinsing is, therefore, essential.

2. Removal of fingerprints
(a) Soap or detergent and water applied with a sponge, rinse with clean water and wipe dry if necessary.
(b) By the use of solvents, apply proprietary solvents according to manufacturer's instructions followed by a thorough rinsing in clean water. Non-proprietary solvents include acetone, petrol, white spirit, naphtha, perchlorethylene, toluol, trichlorethylene.
(c) By the use of solvent mixed with a mild detergent and water. Apply with cloth or sponge, rinse with clean water, wipe dry. The mixture should be stirred before each application. This method may be more effective than using a solvent alone since although solvents remove grease etc., they can leave surface smears.
(d) By the use of mild abrasive cleaners as per item 1b.

3. Grease pencil markings
(a) Soap or detergent as per item 1a.
(b) Solvent as per item 2b.
(c) Solvent mixed with a mild detergent and water as per item 2c.
(d) Proprietary phosphoric acid cleaner. The stainless steel surface should always be rinsed with water before the acid cleaner is applied. This is to wash away any chlorides that would cause etching if present.

4. Lead pencil markings
(a) Mild abrasive cleaner as per item 1b.
(b) Proprietary phosphoric acid cleaner as per item 3d.

These treatments will erase marks but any scratching of the metal surface by the pencil point will not be removed.

5. Removal of plaster
(a) Stronger abrasives, including grinding wheels, impregnated nylon pads, stainless steel wire wool and stainless steel wire brushes. With unidirectional polished finishes application should be in the direction of polishing.
Ordinary steel wire brushes and wool should never be used on stainless steel.

6. Removal of rust and other corrosion products
(a) Mild abrasive cleaner as per item 1b.
(b) Proprietary phosphoric acid cleaner as per item 3d.
(c) Oxalic acid, the cleaning solution should be applied with a swab and allowed to stand for 15–20 minutes before being washed away with water. The stainless steel surface should then be dried with a soft cloth.

7. Removal of scratches
(a) Impregnated nylon pads. Slight scratches on unidirectional polished finishes can be removed. Application should be in the direction of polishing. Scratches cannot be removed on mill finishes. It is difficult to match the original polished finish as supplied by the mills.
(b) Polishing with scurfs dressed with iron-free abrasives, for deeper scratches. Apply in the direction of polishing. Recommended abrasive grit sizes to be used are 80, 120 and 150.
* Generally the stainless steel plates used are 150 grit hand sander finish.

8. Water scale and staining
(a) Wipe with damp cloth.
(b) Mild abrasive cleaners as per item 1b.

(c) Proprietary phosphoric acid cleaner as per item 3d.

Ironmongery The following is recommended by the manufacturer Messrs. Smith Widdowson & Eadem Ltd, 296 Penistone Road Sheffield S6 2FT.

All internal furniture should be cleaned periodically with a cloth damped in warm soapy water then dried and polished with a dry soft cloth.

Locks
All locks are pre-lubricated with a special lock mechanism grease and as such require no specific maintenance.

The following ironmongery has been used generally.
(a) Ref. 9J3 stainless steel butt hinges 100 mm.
(b) Ref. 1G/2 satin anodised aluminium lever handles.
(c) Ref. 0A/1 76 mm mortice lock.
(d) Ref. 0C/1 76 mm mortice latch.
(e) Ref. 1G/5 satin anodised aluminium lever handles.
(f) Ref. 0D/6 rebate set.
(g) Ref. 4B/2 satin anodised aluminium flush bolts.
(h) Ref. 4B/5 socket for concrete.
(i) Ref. 1L/2 satin anodised aluminium lever handles.
(j) Ref. 5A/13 satin anodised aluminium door closers.
(k) Ref. 1L/5 satin anodised aluminium lever handles.
(l) Ref. 2B/2 229 mm satin anodised aluminium pull handles.
(m) Ref. 7D/1 single panic bolt.
(n) Ref. 0A/3 mortice deadlock.
(o) Ref. 1C/A satin anodised aluminium escutcheons.
(p) Ref. 4B/6 easy clean floor socket.
(q) Backplates to pull handles, finger plates, and pictograms are in 10 gauge satin anodised aluminium.
(r) Kicking plates are 16 gauge satin anodised aluminium.
(s) Chubb mortice Bolt Ref. 8001.
(t) Ref. 11A/1, 11A/3, 11A/5, 11A/7 Name plate slides and inserts.
(u) Universal concealed door closer, heavy duty.
(v) Ref. 5C/1 floor mounted door stop.
(w) Ref. 5C/6 wall mounted door stop.

Suite arrangements
All separate office areas are suited differently but all offices on one floor will pass under the sub-master key cut for that floor.
4 sub-master keys are provided per floor.
2 only Grand Master keys are provided and these will open all office areas on any floor.
All duct doors and cleaners stores pass under suite z and y keys respectively.

10.03 Terrazzo paving to Greek Street entrance

Basement doors other than the strong room doors are standard differs.

Floors
Liquid detergents, detergent powders, soap and soap powders whilst not affecting the structure of terrazzo will, over a period of time, build up a thin film which will in itself hold dirt and also make the floor slippery.
Terrazzo should be cleaned with a mildly abrasive cleaner which is chemically inert. This should be applied liberally with plenty of water and the surface well scoured using a scrubbing machine. Care should be taken to remove all traces of the cleaner afterwards to prevent any tendency for the floor to be gritty underfoot. The use of mops should be avoided as this will eventually lead to a build up of dirt which can prove difficult to remove.
The frequency with which a terrazzo floor requires cleaning naturally depends to a large extent upon the nature and density of the traffic it receives. As a general guide a thorough cleaning once per week is sufficient on commercial premises with a light cleaning at intermediate intervals should the floor require it.
Frequent cleaning during the first few months is particularly important as this will help to bring the floor to its best appearance by removing the 'bloom'' which is a characteristic of newly laid terrazzo.

Flexopave screed epoxy resin This material will withstand superimposed loading and the effect of trucking. The floor should be cleaned by washing with detergent and water, but this media is resistant to cleaning with hydrochloric acid or sodium hypochlorite.

Plastylon PVC floor coverings *Primary maintenance*
The floor should be cleaned with water containing a small quantity of commercial detergent. To give the floor a shiny appearance apply two coats of a water emulsion wax by means of a nylon foam brush or other appliance. The second coat should be applied 20 minutes after the first.

General maintenance
Clean the floor with a damp cloth. Should the surface become worn apply another coat of water emulsion wax to the affected area.
To remove scuff marks dust with abrasive powder e.g. Ajax or Vim and then rub the affected area with a cloth dipped in warm water but not containing detergent. The floor should then be well rinsed with clean warm water.
It is advisable to apply two coats of water emulsion wax on the areas treated as described above.

The handover

Mastic asphalt   Messrs. Tunstalls, the sub-contractors responsible for laying the asphalt, recommend the following:

Maintenance of mastic asphalt is not normally required for 50/60 years but, a visual inspection should be made annually, particularly if the asphalt has been trafficked by tradesmen working on the asphalt.

Russum mats   The mats should be periodically removed from the mat wells and loose dirt and dust removed by pressure vacuum or by turning the mat over and beating. The mat should be kept free of excess water.

Carpets   The Technical Advisory Committee of The Federation of British Carpet Manufacturers recommends the following methods for stain removal:

Method A: Using dry cleaning solvents (e.g. carbon tetrachloride or trichlorethylene). Apply the liquid with an absorbent cloth. Do not pour any liquid on to the carpet. Commencing at the outer edge of the stain, rub very gently and work gradually towards the centre. The surface of the absorbent cloth must be changed frequently and re-moistened with the cleaning liquid.

Mop or blot well between each application. When the standing is heavy, absorbent material should be placed beneath the carpet unless it is fitted. The solvents should be used with care on carpet type 1 'Tankard Walkover' which has a latex back.

Method B: Using shampoo or detergent. If a carpet shampoo is not available a non-alkaline liquid detergent may be used. Prepare a cold or lukewarm solution or foam according to the maker's instructions, and preferably add one teaspoonful of white vinegar per pint of solution. It is better to have the solution weaker than the maker's recommendation, rather than stronger. Apply this solution or foam to the stain with a sponge or absorbent cloth and mop or blot off. Repeat this procedure if the method appears to be effective, until the stain is removed or there is no further beneficial action. Finally treat with lukewarm water in a similar manner and leave the pile sloping correctly. For the method of drying, see the notes below headed 'Water'.

Sometimes, when grease marks appear to have been removed completely, they tend to reappear later owing to traces of grease left at the base of the tufts creeping upwards and attracting dirt. A further treatment usually prevents a recurrence. Dry cleaning liquids dissolve grease and oil but do not dissolve water-soluble substances such as salt or sugar. Therefore, Method A is recommended for the removal of greasy substances, followed by Method B if necessary. Similarly, water-soluble substances are treated by Method B and when the carpet is quite dry by Method A if required.

Alcoholic liquids, beers, wines and spirits
Mop thoroughly and apply Method B. Old stains present difficulties, but sponging with methylated spirit reduces the stain.

Where only partial removal has been effected, a 50:50 solution of glycerine and water should be used. Care should be taken to remove the glycerine after the stain has been treated, by rinsing with water, otherwise the treated area will soil.

Battery acids; strong acid
This strong solution of sulphuric acid, if allowed to penetrate, quickly attacks the backing yarns, therefore prompt action is necessary to prevent serious damage. Blot thoroughly and liberally apply a saturated solution of bicarbonate of soda or baking powder. If these are not available, apply borax, ammonia or washing soda using about one tablespoonful per pint of water. The latter treatments may affect the colours but save the carpet.

Beverages: cocoa, coffee, tea, milk and soft drinks
Method B followed by Method A.

Burns
With wool carpets the blackened wool can be removed with scissors and then use Method B. Man-made fibres react differently to burns and it may need expert attention as will extensive burns.

Fruit stains
Method B, followed by methylated spirit if required.

Grass stains
Mop with methylated spirit and follow with Method B.

Ink stains, writing ink
Blot thoroughly to prevent spreading. Fresh stains may be satisfactorily removed by Method B in many cases. Old stains present difficulties and require treatment by an expert.

Copying ink and red ink
Blot and treat with methylated spirit, preferably with a small addition of acetic acid or white vinegar. Apply with an absorbent cloth and quickly blot to avoid spreading. Persist with the treatment and finally use Method B.

Oil and grease
Mop or blot thoroughly or scrape off grease. Follow by Method A and finish with Method B. Bad staining may re-develop later and require a further treatment.

Paint
Treat as for oil and grease. Paints vary widely in their composition and professional assistance may be required.

Rust
Method B. Old stains are difficult to remove and need expert attention.

Saline liquids
Mop and blot, followed by Method B. Rinse and mop several times with water.

Salt
Powdered salt should be thoroughly removed by a vacuum cleaner. Salt, if allowed to remain, may affect the colours and attract moisture which leads to increased dirt retention.

Shoe polish
Scrape off as much as possible, mop and blot. Then use Method B followed by Method A.

Soot
Vacuum lightly and gently, followed by Methods B and A if required.

Starchy foods
Scrape and follow with Method B.

Tar
Treat as for oil and grease (above). Solvent naphtha (flammable) may prove useful in difficult cases.

Urine
Treat as for saline liquids (above).

Water
Where the wetting is not extensive and the water is comparatively clean, mop thoroughly and dry as quickly as possible. If the carpet is not fitted wall-to-wall, support the wetted portion above the floor and assist drying by the aid of coal, gas or electric fires, fans, etc. Dry in the open air if circumstances permit.

See that the carpet pile is laid correctly prior to drying. On no account should the carpet be put into service again until it is quite dry.

If the water is known to be slightly alkaline, as are many well water supplies, it is advisable to sponge with a weak solution of white or ordinary vinegar, using one egg-cupful per gallon of water.

Carpets saturated with water following bursts or leakages from roofs or faulty heating systems should be well mopped and professional cleaners called in as soon as possible. Failure to deal quickly may result in discolouration, bleeding of colours, shrinkage and mildew.

Wax
Scrape away as much as possible. Next hold a hot iron a few inches above the mark and use Method A.

Note: The carpet tiles can be course be replaced if the above methods fail. Refer to Section 8 for type and colour of tiles specified.

This part continues with similar details on wall finishes, ceilings, fixtures and fittings, services, lifts and sanitary fittings. It ends with a one page schedule of regular cleaning instructions.

10.10   ***Regular cleaning instructions (recommended frequency)***

Item	Operation	Recommended frequency
A General		
Ash trays	Empty and dust and replace	Daily
Waste paper baskets		
Litter bins		
B Furniture, etc.,	Dust	Daily
Polished wood	Polish	Fortnightly
Painted wood and metal	Dust	Daily
	Wash	Annually
PVC covered upholstery	Dust	Daily
	Wash	Monthly
Fabric upholstery	Vacuum clean	Monthly
Plastic	Dust	Daily
	Wipe over	Monthly
Mirrors and glass in bookcases, etc.	Dust	Daily
	Wash	Monthly
C Floors		
Asphalt	Sweep and lightly buff	Weekly
Terrazzo	Sweep and wash	Daily
Carpet and mat wells	Vacuum clean	Daily
D Walls (including windows and doors, skirtings)		
Glazed tile	Dust	Weekly
	Wash	Monthly
Fairfaced blockwork/concrete	Brush down	Annually
Gloss paint		
Emulsion paint		
Semi-gloss paint	Dust	Weekly
Flat oil paint	Wash	Annually
Plastic (Wareite, Formica)		
Granite		
PVC – coated fabric		
Polished hardwood	Dust	Weekly
	Revive polish	Annually
E. Ceilings		
Emulsion paint	Dust	Weekly
	Wash	Annually
F Door furniture Metal finishes		
Anodized aluminium	Dust	Daily
Stainless Steel	Clean	Weekly
G Sanitary fittings	Clean	Daily
H Glass	Clean	Weekly (internally) Monthly (externally)

The handover

Also included in Part 2 of the maintenance manual is the following section which details the architectural record drawings handed over with the document. It should be noted that the mechanical engineering contractor provides record drawings with his own maintenance manual – a frequent practice.

One copy each of the following drawings are provided with this manual consisting of main plans and sections.
If any further information is required please contact the Architects Abbey and Hanson Rowe & Partners, Huddersfield.

Drawings

			scale
(21)	01	Revision	1.20
(4–)	01 F	Basement floor plan	1.50
(4–)	02 D	Ground floor plan	1.50
(4–)	03 B	First floor plan	1.50
(4–)	04 A	Second floor plan	1.50
(4–)	05 B	Third floor plan	1.50
(4–)	06 C	Fourth floor plan	1.50
(4–)	07 C	Fifth floor plan	1.50

Drawings relating to Mechanical Installation are included within the maintenance document prepared by Messrs. How Group Ltd.

Part 3 is concerned wholly with mechanical services and was prepared by the mechanical and electrical consultants, F. R. Jenks and Partners. Once more reference is made to the additional information contained in the mechanical engineering contractor's maintenance manual and his as-built drawings.

The Part starts with a general description of the systems installed and their operation and maintenance.

12.02   *General description of the systems*
In the interests of clarity the system will be dealt with separately under the following headings:

1. *LPHW heating installation and boiler plan*
2. *Plenum heating installation to ground and first floor and ventilating plant.*
3. *Toilet extract systems*
4. *Domestic hot-water service*
5. *Mains cold-water service*
6. *Tank cold water downservice*
7. *Gas installation*
8. *Dry riser installation*

12.03   *LPHW heating installation and boiler plant*
The building is heated by means of a conventional gas-fired low pressure hot water heating installation of the open type, comprising 2 No. atmospheric gas fired boilers with conventional flues, duplicate circulating pumps, feed and expansion tank, radiators and cill line heating natural convectors. In addition, the boilers supply water to the DHWS storage calorifiers and the ventilation plant heater batteries. These will be described in the relevant sections.

The boiler plant supplies water at flow and return temperatures of 82 °C. (180 °F) and 71 °C (160 °F) respectively. The flow water passes through a 3-port motorised mixing valve into the pump suction header where duplicate in-line circulating pumps (one duty, one standby) pump the water round the system. The rising mains are located in a service duct adjacent to the electrical switchgear room on each floor. Flow and return connections to these mains are taken off at high level on first to 4th floor inclusive, and a single pipe ring main run from these connections generally around the perimeter of the building within the false ceiling void. The flow and return connections to individual radiators on the 5th to 2nd floor inclusive, drop directly through the floor to connect into the ring main at high level on the floor below. All radiators are fitted with lockshield union valves on the return connection and Danfoss thermostatic valves on the flow side. Generally these valves are fitted with the in-built sensing heads, except for a few on the 5th floor which have the remote sensors. Radiators are the only form of heating on the 5th floor to 2nd floors inclusive.

Both ground and first floors are supplied with a perimeter cill-line natural convector system to offset the fabric heat losses of the building, and a plenum heating system to cater for the ventilation requirements. The plenum system is dealt with under Section 2.

Generally, the cill-line systems are again fed from a single pipe ring main at high level on the floor below. On the ground floor on the Park Row elevation and parts of both the Greek Street and Russell Street elevations the ring main rises from high level in the basement and runs in a builders work trench cast in the ground floor slab adjacent to the perimeter. Similarly on the first floor on the above mentioned elevations the ring main rises from the ground floor ceiling void, but in this case runs within the convector casings themselves.

Background heating to the basement store rooms is provided by leaving the ring main serving the ground floor unlagged, thus effectively using it as a high level pipe coil.

The controls to the heating system can be sub-divided into two parts i.e.
(a) Boiler controls, (b) System controls.

12.03(a) Boiler controls
Each of the boilers is fitted with its own combined control and limit thermostat, control box and gas train, comprising main gas inlet cock, main gas governor, main gas shut off valve, pilot gas cock and pilot gas control. For a complete description of boiler ignition sequence, please refer to manufacturing literature. The boiler thermostat maintains a boiler flow temperature of 82 °C (180 °F) at all times. Should the boiler water temperature rise above this for any reason the limit thermostat set at 88 °C (190 °F)

will cut off the gas supply and prevent further firing until it has been manually reset.

12.03(b) System controls
A control panel has been installed in the boiler house. This contains all necessary fuses and starters for the 6 No. pumps, fuses for the boiler feeds, control circuit fuse, and start/optimiser programmer. This programmer is connected to the motor on the 3-port valve already referred to, an outside compensator thermostat located on the roof, an immersion thermostat in the main heating pump discharge header and a room thermostat located in the main area on the ground floor. The panel also provides fused feeds for the HWS controls. The 9 No. switches on the face of the panel are as follows:

Main heating pumps duty selector switch	No. 1 – OFF – No. 2
HWS primary pumps duty selector switch	No. 1 – OFF – No. 2
HWS secondary pumps duty selector switch	No. 1 – OFF – No. 2
Boiler No. 1 selector switch	Hand – OFF – Auto
Boiler No. 2 selector switch	Hand – OFF – Auto
Main heating pumps selector switch	Hand – OFF – Auto
HWS primary pumps selector switch	Hand – OFF – Auto
HWS secondary pumps selector switch	Hand – OFF – Auto
Main plant switch	Auto – Night – Day

Under normal operation the pump duty selector switches shall indicate either pump and the remaining 5 No. Hand – Off – Auto switches should be in the 'Auto' position. The optimiser/programmer has been set for a normal working day of 9 hours i.e. 0800 to 1700 hours, with omission on Saturday and Sunday. The programmer correlates the signals from the outside thermostat and the room detector to switch the plant on at such a time as to achieve the pre-determined room temperature at the normal start time. (Note, this applies only to the cill-line and radiator circuit). During the working day the outside thermostat and immersion thermostat in the pump discharge header work in conjunction according to a pre-set schedule to provide flow water to the system at a lower temperature than the boiler temperature, by means of the mixing valve.

Switching to the 'Hand' position on any switch will cause that equipment to run continuously. The 'Day' position on the main plant switch causes the entire plant to run continuously and selection of the 'Night' position brings only the frost protection into circuit.

12.04 Plenum heating installation to ground and first floors
Each of these plants is completely separate (with the exception of the common fresh air inlet and exhaust rising ducts and louvres) and they are identical.

Supply air is drawn from an intake louvre at high level 3rd floor and is drawn down the fresh air duct adjacent to the 2 No. plant rooms. It then passes through a volume control damper, back draught damper, and into the air handling unit which comprises filter section, LPHW, heater battery and centrifugal fan section. On the upstream side of the fan within the confines of the plant room, an attenuator is fitted. The supply air passes through a steel shutter type fire damper and into the distribution ductwork from which it enters the room from ceiling grilles positioned as shown on the drawings. Extract grilles of the same type as the supply allow the air to enter the extract system, which generally parallels the supply system back to the plant room. In the plant room the air passes through a downstream attenuator, axial flow fan, upstream attenuator, back draught damper and volume control damper before reaching the rising main extract duct which discharges it through a louvre at 5th floor level.

LPHW for the heater battery is taken from the rising mains in the duct adjacent to the plant room via a separate circuit. A 3-port motorised diverting valve is installed on the return line from the heater battery.

Each plant room is supplied with a control panel which contains the supply and extract fan starters, all necessary fuses and relays and a time switch. The 3 No. switches on the face of the panel are as follows:

Main plant switch	AUTO – OFF – DAY
Supply fan switch	AUTO – OFF – HAND
Extract fan switch	AUTO – OFF – HAND

Under normal operation at a pre-determined time, the time switch switches on the supply and extract fans provided that hot water is available at the heater battery, so that the fan hold off thermostat is made. A duct detector upstream of the fan controls the motor of the 3-port valve so as to maintain a constant supply air temperature. Switching the 2 No. fan switches to the 'Hand' position will allow either of them to run continuously. Switching the main plant switch to the 'Day' position will allow all the plant to run continuously

The handover

regardless of the settings on the time switch.

12.05 Toilet extract system
Both male and female toilets on the ground to 4th floors inclusive are catered for by the main toilet extract system. The toilets on the 5th floor are dealt with individually.

A belt-driven twin fan Nu-Aire extract unit mounted at the 4th floor roof level is connected to a system of ductwork which rises from high level ground floor in a builders work void within the female toilets. Steel shutter fire dampers are installed within the rising duct at the intermediate slab levels. A common branch duct incorporating a volume control damper centre is connected to the rising duct in the toilet false ceiling void. The branch duct then splits to each toilet area and extract grilles in the ceilings in the toilet areas are connected in via flexible ductwork.

An automatic change over starter panel controlling the twin fan is mounted in the lobby of the female toilets on the 4th floor.

The male toilets on the 5th floor are served by a small duct mounted twin-fan unit located in the void behind the toilet wall discharging through a louvre to outside. This fan is wired through the light switch.

The female toilets on the 5th floor are ventilated by means of a window mounted Vent Axia fan with shutters. This is controlled from a local switch.

12.06 Domestic hot-water service
A central plant DHWS has been installed to provide hot water for the toilet areas and the cleaners sinks.

2 No. 800 litres vertical indirect storage cylinders are mounted at high level in the boiler house. These are cross-connected, but with facility for individual isolation.

Primary LPHW is taken from the boiler flow header and pumped by separate duplicate HWS primary pumps through the coils within the cylinders. It is then returned to the boiler return header. Each cylinder has a 3-port motorised diverting valve located in the return line controlled by separate control and high limit thermostats mounted in the cylinder shutter which keep the contents at the storage temperature. Two additional thermostats are switched into circuit and perform an identical function when the immersion heaters (fitted in the bottom of the cylinders) are in use during the summer months when the heating installation is not in use.

The secondary water rises from the cylinders passes through 3-way vent cocks and then combines and rises up the building in the pipe duct in the toilets (female toilets, ground and first, male toilets 2nd–4th). Connections are taken off at high level, split and then rise up through the floor to feed the ranges in the separate toilets. A return line to provide circulation is provided, which rises up the pipe duct alongside the floor. The DHWS system in the main toilets is thus based on the reversed return principle.

At high level 4th floor, the return traverses the building to the cleaners cupboard and then drops through the building in this position back to the boiler house feeding the sinks on the way. Duplicate HWS secondary pumps are provided on the return, to provide circulation.

The existing incoming main has been retained, but a meter provided in the meter cupboard (under stairs from ground to basement, Russell Street elevation). The water main then splits and rises up the building in two positions. One riser runs in the pipe duct in the toilets to feed the 3 No. new cold water storage tanks at 5th floor slab level and the 2 No. existing cold water storage tanks in the 5th floor ceiling void. At each level a connection is taken in the 5th floor ceiling void. At each level a connection is taken from the rising main to feed the drinking fountains on each floor. The second main rises in the cleaners cupboards feeding the sinks there, to high level 5th floor where it feeds the heating f. and e. tank in the ceiling void outside the cleaners cupboards. A stop cock and valve chamber have been provided immediately outside the building line in Russell Street in accordance with Water Board requirements to enable the supply to the building to be isolated.

12.07 Tank cold-water downservice
The tank cold-water downservice is divided into two separate parts, i.e. that which is connected to the new cold water tanks, and that connected to the existing slate tanks.

12.08a New tank installation
The 3 No. new galvanised cold water storage tanks are positioned in a void adjacent to the 5th floor female toilets, and are mounted at 5th floor slab level. The tanks are cross-connected and a single downservice main drops in the pipe duct in the toilet area to serve the urinals and w.c.s only. The urinals 2nd floor to 4th floor inclusive are fed from a connection taken from high level and run in the false ceiling void. The feeds to the w.c.s follow the route of the DHWS i.e. rise in the individual toilets from high level on the floor below. On ground and first floors both urinals and w.c.s are fed from the same connection taken from the main and run at low level behind the toilet panelling.

12.08b Existing tank installation
The new downservice from the existing

FLOOR PLANS

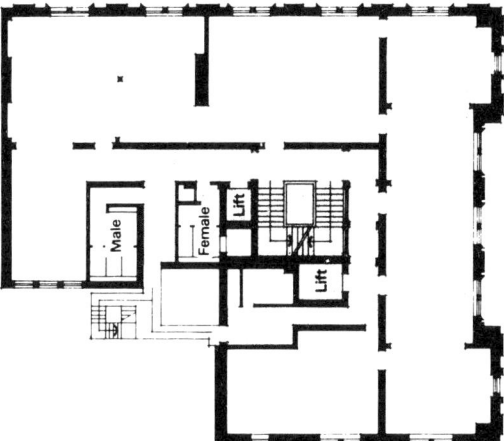

TYPICAL FLOOR

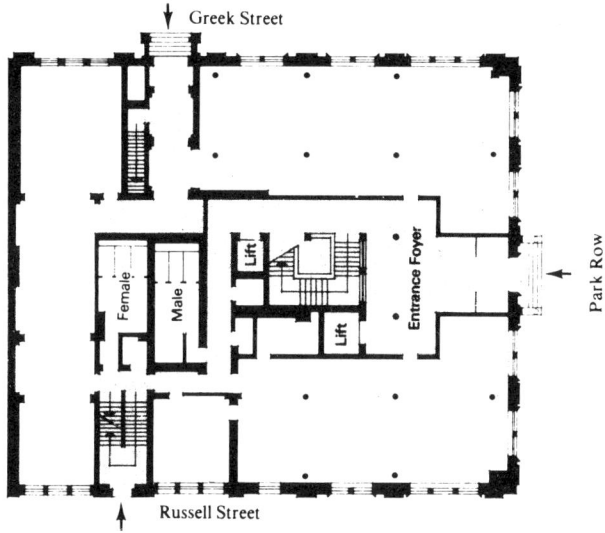

GROUND FLOOR

Figure 7.1 Letting agent's brochure; Lloyds Bank Building, Leeds (courtesy Adair Davy & Mosley)

slate tanks feeds all cold water points on the 5th floor and the spray taps only on the remaining floors, with the exception of the invalid toilets on the ground floor whose w.c. is also connected to this system. Pipework runs generally follow the same route on the new tank installation pipework.

12.09 *Gas installation*
A new main has been run from the existing service in Park Row beneath the Russell Street pavement to the entry point to the building (which is adjacent to the water entry point). The local Gas Board (Negas) have supplied a meter complete with by-pass arrangement which is installed in the same room as the water meter. The line from the meter outlet to the boiler house runs at high level in the basement. The gas installation is connected to the 2 No. boiler gas mains only.

12.10 *Dry riser installation*
The dry riser installation consists of a breaching piece mounted in a steel cabinet at ground floor level on the Park Row elevation connected by 4" galvanised tube to 5 No. banding valves and cabinets mounted within the stairwell.

A list of equipment, complete with manufacturer's address and telephone number, brings the maintenance manual to a close.

Estate agent's letting brochure

To complete the story of the refurbishment of 31 to 32 Park Row, the last piece of documentation produced was the letting agent's brochure.

This contained a short description of the communication links between Leeds and the rest of Britain – proximity of motorways, airports and Inter City rail links. It also contained a small block plan of the centre of Leeds and the relationship of the property to car parks, railway stations and major roads.

A typical upper floor plan and the ground floor plan were reproduced (Fig. 7.1) together with a schedule of areas available for tenancy agreements.

Finally the terms of the leases were listed; such points as the length of the leasing periods, rent review intervals, the situation regarding the payments of rates and a service charge to cover costs that were common to all tenants, and finally the tenants' responsibility for internal repair, decoration and landlord's legal costs in the preparation of the lease and stamp duty thereon.

Appendix 1.

Proprietary names and addresses

Drawing refurbishment

Sarat (Process Photography) Ltd
Vale Road, Portslade, Sussex BN4 1GD

Photographic surveys

Plowman Craven and Associates
Grosvenor Buildings, 104–108 London Road, St. Albans, Herts AL1 1NX

Foundation reinforcement

Carrigan Underpin Limited
Northgate House, Town Square, Basildon, Essex SS14 1EA

Fondedile Foundations Limited
192 High Street, Yiewsley, Middx UB7 7BE

Ground Engineering Limited
Manor Way, Borehamwood, Herts WD6 1WH

Dampness

Companies giving anti-damp services or producing damp-proofing products
Cementone-Beaver Ltd
(Wykamol Division), Tingewick Road, Buckingham MK18 1AN

Peter Cox Limited
Wandle Way, Mitcham, Surrey CR4 4NB

Ground Engineering Limited
Manor Way, Borehamwood, Herts WD6 1WH

Thomas Ness Limited
Eastwood Hall, Eastwood, Nottingham NG16 3ED

Remtox (Chemicals) Ltd
Old Brickyard Industrial Estate, Gillingham, Dorset

Norman Rudd Ltd
110 London Road, Aston Clinton, Bucks HP22 5HS

Thoro NV
Freepost, Sevenoaks, Kent TN13 2BR

The Triton Chemical Manufacturing Co. Ltd
Triton House, Lyndean Industrial Estate, Felixstowe Road, London SE2

Moisture meter manufacturers
Protimeter Limited
Meter House, Fieldhouse Lane, Marlow, Bucks SL7 1LX

Wood preservation and care:

Companies giving services or producing materials
Cementone-Beaver Ltd
(Wykamol Division), Tingewick Road, Buckingham MK18 1AN

Peter Cox Ltd
Wandle Way, Mitcham, Surrey CR4 4NB

Hicksons Timber Products Ltd
Castleford, W. Yorks WF10 2JT

Rentokil Limited
East Grinstead, W. Sussex RH19 2JY

Rickards Timber Treatment Ltd
205 Crow Lane, Romford, Essex RM7 0ES

Sovereign Chemical Industries Co. Ltd
Park Road, Barrow-in-Furness, Cumbria

The Triton Chemical Manufacturing Co. Ltd
Triton House, Lyndean Industrial Estate, Felixstowe Road, London SE2

Winn and Coales (Denso) Ltd (Sylglas)
Denso House, Chapel Road, London SE27 0TR

Structural surveys of walls

British Industrial X-rays Ltd
Commerce Way, Leighton Buzzard, Beds

Renofora (UK) Ltd
Convervation House, Darwen Road, Bromley Cross, Bolton

Wall restoration and maintenance

Companies, and product manufacturers
Aerocem Limited
27 Emperor's Gate, London SW7

Alfred Bagnall & Sons Ltd
6 Manor Lane, Shipley, W. Yorks BC18 3RD

Balfour Beatty Power Construction Ltd
PO Box 12, Acornfield Road, Kirkby, Liverpool L33 7UG

J. Bysouth Ltd
Dorset Road, Tottenham, London N15 5AL

Peter Cox Ltd
Wandle Way, Mitcham, Surrey CR4 4NB

R. Fox & Sons Ltd
38–40 St Pancras Way, London NW1 0QP

T. H. Higgins Ltd
Midland Road, Wellingborough, Northants

Inerlol Co. Ltd
PO Box 2, The Brow, Burgess Hill, W. Sussex RH 15 9NE

Lloyds of Bedwyn
Great Bedwyn, Marleborough, Wilts

London Stone Cleaning and Restoration Ltd
1/3 Miles Street, London SW8 1RX

Natural Stone Quarries Ltd
Springwell Quarries, Springwell, Gateshead, Tyne and Wear NE9 7SQ

Neolith Chemicals Limited
Peel Mill, Market Street, Rochdale, Lancs

New Stone and Restoration Ltd
1 Pembroke Road, Ruislip, Middx HA4 8NQ

Solignum Limited
Thames Road, Crayford, Dartford, Kent DA1 4QJ

Stone Firms (Restoration) Ltd
20 Manvers Street, Bath BA1 1LX

Stoneguard Ltd
The Estate Office, Highgrove Way, Eastcote Road, Ruislip, Middx

Stuarts Granolithic Co. Ltd
46 Duff Street, Edinburgh EH11 2HP (also at Harrow, Stockport and Birmingham)

Szerelmey (UK) Ltd
371/375 Kennington Lane, London SE11 5RA

W. and J. R. Watson Ltd
Romano House, Station Road, Corstophine, Edinburgh EH12 7AQ

Cavity wall tie replacement
Product manufacturers
Harris and Edgar Ltd
Progress Works, 222 Purley Way, Croydon CR9 4JH

Hilti (GB) Ltd
Hilti House, Chester Road, Manchester M16 0GW

GRC makers
Glass Reinforced Concrete (GRC) Ltd
Wincham Lane, Wincham, Northwich, Cheshire CW9 6DE

Faience and terra cotta
Hathernware Ceramic Ltd
Hathern Station Works, Loughborough, Leics LE12 5EW

External wall insulation
Blue Circle Industries Ltd
Portland House, Stag Place, London SW1E 5BJ

Cape Insulation Services Limited
Rosanne House, Bridge Road, Welwyn Garden City, Herts AL8 6UE

Disbotherm Ltd
12 Mount Ephraim Road, Tunbridge Wells, Kent TN1 1EE

Tinsley Wire (Sheffield) Ltd
PO Box 119, Shepcote Lane, Sheffield S9 1TY

Joint re-sealing
Evode Joint Sealing Ltd
Common Road, Stafford ST16 3EH

Tremco Ltd
Key House, Horton Road, West Drayton, Middx UB7 8HP

Glass and glazing products

(Including film and secondary glazing systems)
Alcan Safety Glass Ltd
Knowsthorpe Gate, Cross Green Industrial Estate, Leeds LS9 0NS

Doulton Tempered Glass Ltd
Ripley Road, Bradford, W. Yorks BD4 7TP

Expanded Metal Co. Ltd
PO Box 14, Longhill Industrial Estate (North), Hartlepool TS25 1PR

HAT Glass-Pearson Ltd
PO Box 48, Orgreave Drive, Dore House Industrial Estate, Sheffield S13 9NU

Klingshield Limited
751 Barking Road, Plaistow, London E13 9ER

Madico
3 Grundey Street, Hazel Grove, Stockport SK7 4DD

Pilkington Bros Ltd
Prescot Road, St Helens, Merseyside WA10 3TT

Selectaglaze Ltd
Sutton Road, St. Albans, Herts AL1 5LS

3M United Kingdom Ltd
3M House, PO Box 1, Bracknell Berks RD12 1JU

Proprietary names and addresses

Triplex Safety Glass Co. Ltd
Special Products Division, Eckersall Road, Kings Norton Birmingham B38 8SR

UBM Glass Ltd
Glass Works, Dartmouth Street, Birmingham B7 4AP

Stained Glass
Tudorglass Ltd
Central Park Estate, Staines Road, Hounslow, Middx

Thermal curtains and blinds

C. Nathan & Co.
(Verosol) 24 Lisson Grove, London NW1 6UR

RMC Panel Products (Thermoblind) Ltd
Waldorf Way, Denby Dale Road, Wakefield WF2 8HD

Draughtproofing products

Manufacturers and services
Cape Insulation Services Ltd
Rosanne House, Bridge Road, Welwyn Garden City, Herts AL8 6UE

Duraflex Housecraft Ltd
Kingsditch Lane, Cheltenham, Glos GL51 9PD

Kingdom Marketing Co. Ltd
Shapwick Road, Hamworthy, Poole, Dorset BH15 4AP

Schlegel (UK) Engineering Ltd
Henlow Industrial Estate, Henlow Camp, Beds SG16 6DS

Varnamo Rubber Co. (UK) Ltd
15a Bucklersbury, Hitchin, Herts SG5 1BB

Historic window replacement

Cotswold Casement (Engineers) Ltd
Fielding Works, London Road, Moreton-in-Marsh, Glos.

Roof defect diagnosis service

Tremco Ltd
Key House, Horton Road, West Drayton, Middx UB7 8HP

Roof renovation
Repair and relevant product manufacturers

Associated Lead Manufacturers Ltd
Rugby Wharf, Ferry Street, London E14

Colas Products Ltd
(Monoform) Galvin Road, Slough, Berks SL1 4DL

Dynamit Nobel (UK) Ltd
(Trocal) Gateway House, 302/308 High Street, Slough, Berks SL1 1HF

Evode Roofing Ltd
(Tekurat) Common Road, Stafford ST16 3EH

Floorlife-Andek
(RAC Rooftex) 59 Lansdowne Place, Hove, Sussex BN3 1JB

IMI Broderick Structures
Forsyth Road, Sheerwater, Woking, Surrey GU21 5RR

IMCCO-LINE
Imcco House, Cardiff Road, Luton, Beds LU1 1PP

Metra Non-ferrous Metals Ltd
(Metizinc) Pindar Road, Hoddesdon, Herts EN11 0DE

Plaschem Ltd
(Aerodeck) Morris Street, Dumers Lane, Radcliffe, Manchester M26 9GF

Robseal Ltd
Robseal House, 75–87 Eastcourt Avenue, Earley, Reading, Berks

Tunnerised Roofing Co.
96–104 Old Kent Road, London SE1 4NY

Varnamo Rubber Co.(UK) Ltd
15a Bucklersbury, Hitchin, Herts SG5 1BB

Vel-Va-Lube Co. (Holdings) Ltd
Phoenix Street Sutton-in-Ashfield, Notts NG17 4HL

Over-sheeting
Cape Universal Claddings Ltd
PO Box 165, Tolpits, Watford WD1 8QZ

Robseal Ltd
(Retrofit) Robseal House, 75–87 Eastcourt Avenue, Earley, Reading, Berks

TAC Construction Material Ltd
(Overclad) PO Box 22, Trafford Park, Manchester M17 1RU

Thermal rooflights
Williaam Cox Ltd
London Road, Tring, Herts HP23 6HB

Thermal glazing
Doulton Glass Insulation Ltd
(Thermascrene) Wood Road, West Bowling, Bradford, W. Yorks BD5 7TP

Thermocell Ltd
Stonebow House, The Stonebow, York Y01 2NP

Spray-on insulation
Brown (Urecoat)
Moor Lane, Preston PR1 1JQ

Robseal Ltd
(Roofoam) Robseal House, 75–87 Eastcourt Avenue, Earley, Reading, Berks

Ruberoid Insulation Services
(Rubersil) 31 Longwood, Trafford Park, Manchester M17 INP

Solar reflective paint
Blue Circle Industries Ltd
(Roof paint) Portland House, Stag Place, London SW1E 5BJ

Screeton Paintmaker, (Solaflect)
Besson Street, New Cross Gate, London SE14 5AX

Vel-Va-Lube Co. (Holdings) Ltd
(ASP Solar Coat) Phoenix Street Sutton-in-Ashfield, Notts NG17 4HL

Alternative flashings
British Uralite Ltd, (Nuralite)
Higham Works, Higham, Rochester, Kent ME3 7JA

Marley Waterproofing Products, (Sealtite)
PO Box 17, Otford, Sevenoaks, Kent TN14 5EW

Metra Non-ferrous Metals Ltd, (Metiflash)
Pindar Road, Hoddesdon, Herts EN11 0DE

Ruberoid Building Products, (Ruberflash)
Stockingswater Lane, Brimsdown, Enfield, Middx
EN3 7PP

Winn and Coales (Denso) Ltd, (Denso Dampbond)
Denso House, Chapel Road, London SE27 0TR

Lightweight roofings
Eternit Building Products Ltd
Meldreth, Nr Royston, Herts SG8 5RL

Gränges Essem (UK) Ltd
Leon House, 233 High Street, Croydon, Surrey CR0 9XT

TAC Construction Materials Limited
PO Box 22, Trafford Park, Manchester M17 1RU

Reproduction gutters
Alumusc Ltd
Burton Latimer, Kettering, Northants NN15 5JP

Interior refurbishment companies

Ramchester
68 Rochester Row, London SW1P 1TU

Unilock (Project Interiors International)
176/184 Vauxhall Bridge Road, London SW1V 1DX

Ceilings

Environaire Ltd, (Spanoflex)
44 South Street, Chichester, W. Sussex PO19 1DS

Formwood Ltd
Tufthorn Avenue, Coleford, Glos GL16 8PR

Access floors

H. H. Robertson (UK) Ltd
Cromwell road, Ellesmere Port, South Wirral Merseyside

Unilock Ltd
176/184 Vauxhall Bridge Road, London SW1V 1DX

Fire resistant materials

Cape Boards and Panels Ltd
Iver Lane, Uxbridge, Middx UB8 2JQ

Mann McGowan Fabrications Ltd, (Palosol Fireboard)
2–4 Mount Pleasant Road, Aldershot, Hants

Nullifire Limited
46 Park Road, Kenilworth, Warwickshire CV8 2GF

Shapland and Petter Ltd
(Fireguard) Barnstaple, Devon EX31 2AA

TAC Construction Materials Ltd
PO Box 22, Trafford Park, Manchester M17 1RU

Fire venting

Greenwood Airvac Ltd
PO Box 3, Brookside Industrial Estate, Rustington,
Littlehampton, W. Sussex BN16 3LH

Package plumbing

Schott-Kem
Drummond Street, Stafford ST16 3EL

Mallinson-Denny (Bushboard) Ltd
Princesway, Team Valley, Gateshead, Tyre and Wear
NE11 0US

Interior surface refurbishment

Alltek Coatings (UK) Ltd
(Scandatex) 594 Kingston Road, Raynes Park, London
SW20 8DN

Berger Paints, (Tartaruga)
Freshwater Road, Dagenham, Essex RM8 1RU

Norsk Hydro (UK) Ltd
(Respatex panels) Concord House, The Centre, High
Street, Feltham, Middx TW13 4BG

Polycell Products Ltd
Broadwater Road, Welwyn Garden City, Herts AL7 3AZ

Senco Distribution Ltd
Atone House, 38 Poole Road, Bournemouth, Dorset

Tasso Decor International Ltd
Silwood Road, Ascot, Berks SL5 0QU

Unibond Universal Adhesives Ltd
(CeraFix) Tuscam Way, Industrial Estate, Camberley,
Surrey GU15 3DD

Note: This list does not attempt to be comprehensive but concentrates on those companies and products mentioned in Chapter 5.

Appendix 2

Useful references and organisations

The following is a list of references, some of which have been mentioned in the present text.

BRE Digests
113 *Cleaning external surfaces of buildings*
125 *Colourless treatments for masonry*
139 *Control of lichen, moulds and similar growths*
177 *Decay and conservation of stone masonry*
245 *Rising damp in walls: diagnosis and treatment*

British Standards
1282 : 1975 *Guide to the choice, use and application of wood preservatives*
3452 : 1962 *Copper/chrome water-borne wood preservatives and their application*
3453 : 1962 *Fluoride/arsenate/chromate/dinitrophenol water-borne wood preservatives and their application*
4072 : 1974 *Wood perservation by means of water-borne copper/chrome/arsenic compositions*
5268 : Part 5: 1977 *Preservative treatments for constructional timber*
5839 : Part 1: 1980 *Fire detection and alarm systems in buildings. Code of practice for installation and servicing*

Bidwell, T. G., *Conservation of brick buildings,* Brick Development Association
Blacker, Jacob, *Maintenance manual and job diary,* Building Centre
Eldridge, H. J., *Common defects in buildings,* HMSO
Hicken, N. E., *The dry rot problem,* Rentokil 1973
Mills, Edward D., *Building maintenance and preservation,* Butterworth 1980
Price, C. A., *Decay and preservation of natural building stones,* BRE 1974

Further advice can be obtained from the following organisations:

Agrément Board, P.O. Box 195, Bucknalls Lane, Garston, Herts
Architectural Salvage, c/o Hutton and Rostron, Netley House, Gomshall, nr Guildford, Surrey GU5 9QA
British Standards Institution, 2 Park Street, London W1A 2BS
British Wood Preserving Association, Premier House, 150 Southampton Row, London WC1B 5AL
The Building Conservation Trust, Apartment 39, Hampton Court Palace, East Molesey, Surrey KT8 9BS
Building Research Advisory Service, Building Research Station, Garston, Watford WD2 7JR
Contract Cleaning and Maintenance Association, 142 Strand, London WC2R 1HH
Council of the Ironfoundry Association, 14 Pall Mall, London SW1Y 5LZ
Fire Research Station, BRE Borehamwood, Herts WD6 2BL
The Stone Federation, Admin House, Market Square (North Side), Leighton Buzzard, Beds LU7 7EU
Timber Research and Development Association, Stocking Lane, Hughenden Valley, High Wycombe, Bucks HP14 4AD

Index

Abbey and Hanson Rowe and Partners, 6, 12 *et seq.*, 77 *et seq.*
Abbey House, Glasgow, 46–7
access floor, 62
Adair Davy and Moseley, 12, 85
Aerocem, 45
Aerodeck, 57
Agrément Certificate, 36
Alcan Safety glass, 52
Alexander Court, Wandsworth, 55
Alfa Laval, 51
Andrews (Raymond) and Partners, 29
Annabelle House, Hounslow, 49
annealed glass, 52
anobium punctatum, 39
Arazzi Clinic, 61
Architectural Association, 9
ascomycetes, 40
as-built drawings, 77, 82
asbestos-cement slates, 55
ASP Solar Coat, 56

Balfour Beatty Power Construction, 45
basements, leaking, 37–8
Belhus Park, South Ockendon, 58
Berger, 69
Besse Building, Edmunds Hall, Oxford, 54
bill of quantities, 23
Bituthene, 37
Blue Circle, 49, 56
Bovis Construction, 6, 24, 26, 29, 68, 70, 73
BRE, 36, 39
Brentford Nylons building, 51
Breplasta, 69
British Gypsum, 37
British Industrial X-rays, 43
British Smelter Construction, 61, 68
Brixton Estate's headquarters, 59–60
Broderick Structures, 55–6
Building Design Partnership, 75
Building Regulations, 48, 53, 58, 65
Buroplan 300, 62
butyl roofing, 57

Calder House, W1, 5
Cape Boards and Panels, 61, 65, 66
Cape Insulation Services, 49, 52
Cape Weiflex, 53
carbonation, 49
Carrigan Underpin, 33
Cavendish Hotel, W1, 68
cavity walls, 39, 45
Celcure, 41
cellar fungus, 40
Cemfil, 76
CeruFix, 69
chemical d.p.c., 36
chlorides, 35
Colas Products, 56
Comfort Hotels headquarters, 50–1
condensation, 38, 49
conservation, 4
control systems, 63
cost control document, 23

Index

Cotswold Casements, 53–4
Covent Garden Market, 4, 9, 10, 65
Coxdome rooflights, 58
Cox (Williaam), 58
crack repair, 38, 45–6
Crawford Doors, 53
Curzon Street (31 and 32), 5, 74–5
Cutlers Gardens, 33

dampness, measuring, 33
dampness, rising, 34
death-watch beetle, 39–40
Disbotherm, 49
disposable buildings, 1
District Surveyor, 26
d.p.c., 34–5
d.p.c., inserting, 35 et seq.
draught control, 52–3
dry rot, 39, 40
Dunham–Bush heater, 62
Duraflex, 52

efflorescence, 35, 37, 39
electro-osmosis, 35–6
Embassy House Hotel, Kensington, 64
energy conservation, 62–4
Environaire, 60, 62
EPDM, 52
Esselte headquarters, 68
estimate of prime cost, 23
Evode, 39, 57
Examination School, Oxford, 53

fast-track programme, 27
feasibility study, 12 et seq.
fee construction, 6, 23 et seq.
fire officer, 14, 22
Fire Research Station, 65–7
fire resistance upgrading, 65–7
Flexi-wall, 69
floor covering cleaning, 79–81
Floorlife-Andek, 57
Fondedile Foundations, 33
Formwood, 59, 62
Fowler, 9
fungi imperfecti, 40
furniture beetle, 39

gamma radiographic investigation, 43
Gangnail, 54
GenRad headquarters, 5 et seq.
Glass Reinforced Concrete, 46, 76
GLC, 49, 52
Gränges Essem, 55
gravity grouting, 44–5
GRC, 46, 76
Greenwood Airvac, 67–8
Grosvenor Hotel, Glasgow, 46
Ground Engineering, 33
grouting, 38, 44, 45
Gyproc Thermal board, 49

hand-over manual, 77 et seq.
HAT, 50
Harris and Edgar, 45

Hathenware Ceramics, 47
Hazlitt Theatre, Maidstone, 62
HB Roofbond (East Midlands), 55
heat detectors, 65
heat reclaim, 63
Hemax, 63, 45
Hicksons, 41
high build paint, 69
Hillaldam Coburn, 53
Hilti Wall-tie, 45
Holland, Hannen and Cubitts (Northern), 75
hydraulic lifts, 61
hygroscopic salts, 35, 38
hylotruper bajulus, 40

Icosit, 50
Imcco, 55
indenting, 46
industrialised building, 3, 10
Industrial Revolution, 4
Inertol, 49
information schedule, 23, 28
insulating shutters, 53
insulation, roof, 57
insulation, wall, 48–9
International Modern Movement, 52
intumescent coatings, 65
intumescent seals, 61

Jenks (F.R.) and Partners, 12, 13, 14, 82 et seq.

Kingdom, 52
Kirkwood (Alexander), 46
Korrugal Villa tiles, 55

leaded lights, 53–4
lighting economy, 63
Limpet board, 61, 66
Lincolnshire County Architects, 72
Lloyds Bank Building, Leeds, 6 et seq., 12 et seq., 77 et seq.
Llamar, 50
longhorn beetle, 39–40
loose-lay roofing, 56–7
lyctidae 40
Lyl and Scott of Hawick, 58

Madico, 50, 52
maintenance, 3, 77 et seq.
maintenance manual, 77 et seq.
Maltings, Snape, 10, 73
Marley Waterproofing Products, 55
Marshall Engineering, 55
Matcham (Frank), 68, 70
merulius lacrymans, 39, 40
Metallack, 55
Metizinc, 55
Metra Non-ferrous Metals, 55
Midshire Building Society headquarters, 59
Miller (T.M.) and Partners, 46
mine fungus, 40
moisture meter, 33–4
Monoform, 56
Monolux, 66
Multi-glass Antisun units, 50
Murdock Design Associates, 5, 6, 24

NAFTA offices, 59
Neolith, 48
nitrates, 35
Norman Rudd Freezteq, 36
Norsk Hydro, 69
Norton Priory, Runcorn, 54
Nullifire, 65

Old Brew House, Castle Bromwich, 36
Opera House, Buxton, 68
operational document, 23, 24 et seq.
organic solvent preservation, 41
Ove Arup and Partners, 12 et seq., 70, 73
Overclad, 56
Oxford Street (527–531), 23, 29
Oxford University 53

Palesit mortar, 50
Pali Radice, 33
Palusol Fireboard, 61
partition performance, 58–9, 66
Peroglaze, 49
Peter Cox, 36
PFA, 46
Phipps (C.J.), 70
photomanual drawing restoration, 33
Pilkington Brothers, 50, 52, 75
Pilkington Training Centre, 11, 75–6
planning officer, 14
Plascem, 57
plasterboard, 58–9
plaster-in-a-roll, 69
plastic glazing, 52
plastic repairs, 46
plusaire roof ventilators, 67–8
Pluscarden Abbey, 54
Pluvex Ruberflash, 55
Polevo, 39
Polycell, 69
powder post beetle, 40
pozzolanic additives, 46
prefabricated plumbing units, 64
preservative, organic solvent, 41
preservative, tar oil, 41
preservative, waterborne, 41
pressure grouting, 45
Prevac, 41
Pride (F.H.), 62
Project Interiors International, 29, 60
Protim, 41
Protimeter, 34
Public Health Act, 34
Pudlo, 48
PVC membrane, 56–7

Quickpile system, 33

RAC Rooftex, 57
Railways, Shops and Offices Act, 15
rain penetration, 38–9
raised service floor, 62
Ramchester, 29
record drawings, 13, 31
Reflecto-Shield, 50
relative humidity, 35
Renofora, 43

Rentokil, 41
Renton Howard Wood Levin Partnership, 70
replacement windows, 50
repointing, 46
Respatex panels, 69
resurfacing, 46
Retrofil system, 56
Rickards Timber Treatment, 42–4
RMC Panel Products, 53
Robertson (H.H.), 62
Robseal, 56–7
Rocksil, 49
rooflights, 58
rot, dry, 40
Rotunda, Birmingham, 39
rot, wet, 40
Royal Liver Building, Liverpool, 73
Ruberoid Building Products, 55

salts analysis, 35
Sam Newsom Music Centre, 10, 72–3
Sarat (Process Photography), 33
Schlegel, 52
Schott-Kem, 64
Screeton Paintmaker, 56
Scotchtint, 50–1
Sealtite, 55
secondary glazing, 51
Senso Distributors, 69
SGB, 5
shut-offs (intumescent), 61
Smith (W.H.), Worthing, 67
smoke detectors, 64–5
Société Cysla d'Industrie, 29
soft rot fungus, 40
Solaflect, 56
Sound Research Laboratories, Sudbury, 51
Sovereign Deepkill, 41
Spa-Line units, 64
Spanoflex, 60–1, 62
spray plaster, 69
stained glass, 54
stainless steel cleaning, 78
Staple Hall, 61
Staple Inn Hall, Holborn, 42
Statutory Approvals 23, 26, 29
Stephen Y, 23 et seq., 29
St Mary's Church, Winkfield, 42
St Mary the Virgin, Woodham Ferrers, 41
stone cleaing, 48
Stone Federation (The), 46
Styrocote, 49
sulphates, 35
Sungard, 50
Supalux board, 61, 65–6
survey techniques, 31
suspended ceilings, 59–61
Sutton Place, Surrey, 47
Sylglas, 41
Szerelmey, 48

TAC, 56, 61, 65
Tanolith, 41
Tartagura, 69
Tasso, 68
Tekurat, 57
tempered glass, 52
Thaltox Q, 48

Index

Thames Tunnel Mills, 5
Theatre Royal, Nottingham, 10, 70–2
Thermacote, 49
Thermalath, 49
thermal insulation, 48–9
Thermascrene, 58
Therm-O, 65
Thermoblind, 53
Thermocell, 58
Thermocore, 54
Thermorend, 49
Tinsley Wire, 49
TRADA, 66
Treatim, 41
Tremco, 54
Triplex Safety Glass, 52
Tudorglass, 54
Turnerised Roofing Company, 55

UBM Glass, 52
ultrasonic testing, 43
Unibond, 69
Unilock, 59, 60, 62
Urecoat, 56

Vac-Vac, 41
Vapour barrier, 37, 57
Varnamo Rubber, 52, 57
Vel-Va-Lube, 56
Verosol, 50–1
Victoria and Albert Museum, 47

Wallace (Robert), 72
wall cleaning, 48
wall lining, 61
Wareham Town Hall, 55
Warley Hospital, Brentwood, 69
Waterhouse, Alfred, 6
water repellents, 39, 48, 77–8
wet rot, 40
Whitworth's Produce, 56
wired glass, 52

xestobium rufovillosum, 40

Yeomans and Partners, 50, 71, 74–5